A LITTLE
QUEER
NATURAL
HISTORY

Josh L. Davis

The University of Chicago Press

The University of Chicago Press,
Chicago 60637

Published 2024

Printed in China

33 32 31 30 29 28 27 26 25 24   1 2 3 4 5

ISBN-13: 978-0-226-83703-1 (paper)
ISBN-13: 978-0-226-83682-9 (e-book)

DOI: https://doi.org/10.7208/
chicago/9780226836829.001.0001

First published as *A Little Gay Natural
History* in 2024 by the Natural History
Museum, Cromwell Road, London
SW7 5BD

Library of Congress Control Number:
2023951075

Designed by Mercer Design, London
Reproduction by Saxon Digital Services, UK
Printing by Toppan Leefung Printing Ltd.,
China

The origins of this book can be found at the beginning of 2019, not long after I'd started working at the Natural History Museum, when I helped co-develop and lead the Museum's first-ever LGBTQ+ natural history tours. Following on from this, the tours went on to become an award-nominated YouTube series (thank you Lizzie!) and now, a book. For this, I am utterly indebted to the ever-talented Florence Okoye, a fantastic friend who put me on this path when she first asked if I'd like to help her with those tours all those years ago. Thank you!

It would be remiss of me not to say a massive thank you to my long-suffering partner, Chris, who has put up with me recounting everything I've learned in detail and on repeat for months on end, and a special mention to (the late) Bramwell the cat who, despite having a shaky grasp on the English language, made some excellent contributions.

There are a number of incredible scientists who have all generously given up their free time to read, fact-check and comment on each section of this book. Without them none of this would have been even remotely possible. In no particular order (before anyone gets suspicious), thank you so much to Dr Natalie Cooper for her mammalian knowhow, James Maclaine and Dr Matthew Gollock for their fishy knowledge and Dr Gavin Broad and Dr Erica McAlister for representing the invertebrates. Thank you to Dr Alex Bond and Dr Martin Stervander for their avian wisdom, Dr Jeff Streicher and Dr Ian Brennan for their herpetological help, Professor Paul Barrett for covering the dinosaurs and Dr Sandy Knapp, Dr Gothamie Weerakoon and Dr Anne Jungblut for their indomitable botanical and fungal expertise. Finally, a massive thank you to Dr Ross Brooks, who has provided historical knowledge, encouragement and support whenever I needed it. All of these people are world experts in their fields with decades of experience behind them, and I cannot be more grateful for their help and time with this book. Needless to say, this book would not exist without the fine folk in the Museum's Publishing team; a well-deserved thank-you to them all for backing this book and for their incredible patience and proficiency as they guided me through the entire process.

Finally! Just a huge thank you to all the queer folk out there! You might even be reading this right now, and if so, you're amazing and don't let anyone tell you otherwise.

Contents

Introduction

The planet on which we live is filled with an extraordinary range of animals, plants and fungi. Collectively, they exhibit an astonishing diversity when it comes to what they look like, where they live and how they behave. And nowhere is this truer than when it comes to their sex and sexual behaviours.

It is often quoted that around 1,500 species of animals exhibit some form of homosexual behaviour. This includes animals from right across the tree of life: Hawaiian orb weaver spiders and common slipper shells, house flies, nematode worms and Humbolt squid, wood turtles, blackstripe topminnows, Guianan cock-of-the-rocks and brown bears. But this figure is likely a massive underestimate. Considering these behaviours are found in almost every branch of the evolutionary tree, it seems highly unlikely that they are limited to just a few hundred species out of some 2.13 million named to date. The most obvious assumption is that most species of animal probably exhibit some form of queer behaviour, and being a purely heterosexual species is the exception.

But, despite this likely commonality of non-heterosexual behaviour, for most of the history of science it has been covered up, ignored or disparaged. For example, for a long time it was often assumed that animals only have sex to reproduce, and therefore homosexual behaviour was an evolutionary 'paradox'. But it is now known that this is not necessarily the case. Sex is a complex behaviour that can – and does – have multiple causes and outcomes, from stress relief, cementing bonds to just pure pleasure.

It might be useful to start by setting out what this book is not. It is not a justification for queer behaviours – animal or otherwise – as these behaviours need

Queer behaviours have been observed right across the tree of life, including in the Guianan cock-of-the-rock (*Rupicola rupicola*) where around 40% of males engage in some form of homosexual behaviour.

no justification. It is not a comprehensive list of queer behaviours of identities, and it is not making analogies between what is seen in nature and the human race. While humans are part of the natural world, and some of what is discussed here might be applicable when it comes to biases, people have identities which are difficult to discern in non-human species.

The personal selection of examples in this book aims simply to give an overview of the sheer diversity of non-heteronormative biology and behaviours that exists in the natural world – those that are not based on the assumed binary of males and females taking on the 'traditional' roles that have historically been presented.

When it comes to talking about sex and sexuality in nature, it is important to be clear about the language being used. Hopefully some of the examples used in this book help to show how language has played a significant role when it comes to the subject of queer natural history. While not everyone might agree with specific definitions or might question how some words are applied to non-human life, we can at least be clear and consistent with how they are used here.

The first thing to clear up is the difference between sex and gender. For a long time these words have been used interchangeably, even in the scientific literature. Put very simplistically, sex is typically defined as referring to physical, hormonal and genetic characteristics, while gender to an innate sense of self. It is not possible to know if animals have a sense of gender, but many scientists and science writers have, all the same, used the term incorrectly when they are actually talking about sex.

Defining sex itself is also not that straightforward, as it can be dependent on a number of different characteristics, including but not limited to genetics, hormones and anatomy, and is a spectrum rather than clear cut categories. For the purpose of this book it is defined by the size of the sex cell produced by an individual. This means that 'female' is used to refer to the individuals of a species that typically produce larger sex cells, while 'male' is used to refer to those that typically produce smaller sex cells. This book sets out to illustrate a few different ways in which the natural world challenges using single characteristics on which to base complex definitions such as sex. A good example of this is the wholesale application of the sex categories 'male' and 'female' onto plants. While this is common practice amongst botanists, it is also understood that in many cases the main body of the individual plant is often technically asexual in nature, due to something known as 'the alternation of generations'. During this process, the main plant and its flower do not actually produce sex cells, but rather asexual spores which then go on to produce what is known as a 'gametophyte', which itself creates the sex cells. In some species, the gametophyte is even a separate, independent organism.

This might sound like splitting hairs, but if we try to apply terms such as 'male' and 'female' onto these organisms, to what, exactly, are we then referring – the main body of the plant which, technically, is not producing any sex cells itself, or the gametophyte, which may or may not be an independent organism? In order to try and keep things as consistent as possible, this book uses 'female' and 'male' when talking about plants, their flowers and the structures they contain, although with the understanding that, as ever, things are more complex than they may first appear.

Talking now about sexual behaviour, terms such as 'gay', 'lesbian', 'LGBTQ+' and 'queer' are typically restricted to humans, as in many cases they refer to more than just behaviour and, like gender, rely on an innate sense of self. In this context, the word 'gay' is also able to refer to any individual who engages in same-sex behaviour or relationships, meaning both men and women can be described as gay. But it is impossible to ask other lifeforms if they are actually gay, lesbian, bisexual or queer.

Instead, the scientific literature uses a variety of terms to describe these behaviours, usually something along the lines of 'same-sex sexual behaviour'. But, as this term is rather clunky this book instead uses terms such as 'gay behaviour' or 'lesbian activity' (the key here being the inclusion of the words 'behaviour' and 'activity'). To try and make things as consistent and as easy as possible to follow, this book uses the term 'gay' to refer to male-male typical behaviours, 'lesbian' when talking about female-female behaviours, 'homosexual' when referring to either and 'queer' as something of an umbrella term when talking about more general non-heteronormative behaviours that are not necessarily related to sexual behaviour alone.

I hope that this book and the examples it highlights flip the question of 'why does homosexuality exist in nature when it appears to go against evolution?' on its head. Instead, the majority of animals – and many of the plants – are out there right now, most likely engaging in plenty of queer activities without us even knowing.

Adélie penguin
Homosexual couples

Around 20 species of penguin are found throughout the southern hemisphere, with only the Galápagos penguin (*Spheniscus mendiculus*) making it as far as the equator and on to the Galápagos Islands. Famous for their smart attire and inelegant saunter, southern hemisphere penguins have also become known in more recent times for another reason: their queer behaviour. This was popularized in 2005 by the successful children's book *And Tango Makes Three*, the real-life tale of a pair of male chinstrap penguins (*Pygoscelis antarcticus*) called Roy and Silo in New York's Central Park Zoo that paired up and successfully raised a chick together. Yet Roy and Silo are certainly not alone among their kind.

Homosexual penguin couples, both male and female, have been reported from around at least a third of all species of the sartorial birds. While frequently reported in easier-to-observe captive animals, it is also widely described for their wild counterparts. Scientists have been observing queer penguins in the wild for more than a hundred years, one of the most famous accounts emanating from the British Antarctic Expedition between 1910 and 1913.

On 15 June 1910 the expedition set off on board the *Terra Nova* from Cardiff, UK, bound for Antarctica. Led by Robert Falcon Scott, the expedition had both scientific and geographic aims, including that fatal race to the South Pole. On board the ship were 65 men, including ship surgeon Dr George Murray Levick. As was usual for this period of exploration, in addition to his duties as the doctor, Levick was also one of the crew's photographers and the zoologist tasked with recording and collecting the animals and plants encountered during the expedition. When the ship arrived in Antarctica, Levick and five other men formed what would become the Northern Party. While Captain Scott and his team raced south to the Pole, Levick and his colleagues sailed north to spend a year studying the geology and wildlife at Cape

Male and female Adélie penguins are very similar in size with similar features, which can make it hard to tell the difference between them.

Adare. There they observed what is now known to be the largest breeding colony of Adélie penguins (*Pygoscelis adeliae*) in the world, which is today thought to contain around 330,000 birds.

Arriving on 17 February 1911, the team spent the next 11 months at Cape Adare, carrying out some of the most important investigations of the entire *Terra Nova* expedition. They recorded not only the wildlife of the region, collecting some 400 species new to science, but also made crucial measurements of its geology, meteorology and glaciology. They were the first people to ever study glaciers outside of Europe and collected critical baseline data, which is now proving useful for climate change studies. During this time, Levick spent three months living amongst the penguin colony and became the first person to witness an entire Adélie penguin breeding season. He took meticulous records of his day-to-day scientific observations on the penguins, seals and skuas. This included behaviour that Levick seemed unable to

 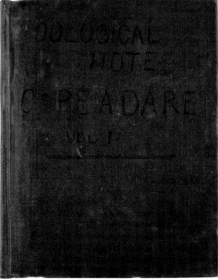

Photographs of Adélie penguins from Scott's expedition to the Antarctic in 1910. It was on this expedition that gay penguins were first observed.

make sense of, including males engaging in necrophilia with dead females, forced copulation and gay sex. Of this, Levick wrote about his confusion: 'Here on one occasion I saw what I took to be a cock copulating with a hen. When he had finished, however, and got off, the apparent hen turned out to be a cock, and the act was again performed with their positions reversed, the original "hen" climbing on to the back of the original cock, whereupon the nature of their proceeding was disclosed.'

This inability to understand or process what he was watching is seemingly reflected in his notes. He referred to the male penguins as 'hooligans' and while some of these behaviours he recorded in English, others were jotted down in a Greek cipher, presumably to obscure them. Even though Levick did not understand what he had seen among the penguins, he tried to publish what he had witnessed. On his return to the UK, he wrote up his notes into a book titled the *Natural history of the Adélie penguin*, including in it a section on the 'hooligan' birds' sexual habits. But it wasn't to be. Before it went to print in 1915, the section on sex was excised from the book by Sidney Frederick Harmer, the then Keeper of Zoology at the Natural History Museum, London, who simply wrote: 'Sexual habits. We will have this cut out and some copies printed for our own use. How many should we want?'

What resulted was 100 copies of *The sexual habits of the Adélie penguin* being privately printed and distributed among a select group who were allowed to see it. Each copy was printed with a bold header reading 'Not for Publication'. Clearly, talking about the behaviours observed in Adélie penguins – including homosexuality – was not deemed acceptable or as something for public understanding. And so, on this occasion, far from being objective, science and the knowledge it creates was regulated and edited. It was not until a hundred years after Levick's observations that these behaviours were reported in full to the public when, in 2012, a copy of this secret pamphlet, discovered sandwiched between the pages of another book, was finally published.

Mangrove killifish
Reproducing with itself

A tiny fish that typically lives in brackish and marine waters along the coast of the Caribbean, the mangrove killifish (*Kryptolebias marmoratus*), might not look like much. Originally described from specimens collected in Cuba in 1880, the species has been found living in swamps and mangroves as far north as Florida, along the coast of Mexico, down into Central and South America to the mouth of the Amazon River, as well as east to the Antilles and Bahamas. Despite this wide distribution the fish is frequently described as 'rare', although this is more likely a result of the fish's unusual living arrangements than a reflection of its population status. This hardy little creature lives in crab burrows, in water in the bottom of discarded cans and even within rotten logs – hardly typical environments for a fish.

Barely reaching 4 centimetres (1½ inches) long, this small fry is notable for a number of reasons. Its extraordinary ability to breathe through its skin means it can survive on land for up to two months, while it can also tolerate extreme water temperatures and hypersaline environments. But it's the mangrove killifish's ability to self-fertilize internally that sets it apart from every other animal with a backbone.

While many species are known to be hermaphrodites – meaning they can produce both male and female sex cells – they still need to mate with other hermaphrodites. The mangrove killifish, along with the closely related *Kryptolebias hermaphroditus*, is unique amongst vertebrates in that they can produce both sperm and eggs and fertilize them within their own body through an act known as 'selfing'.

Its ability to do this is down to a type of tissue known as ovotestis. This tissue is not limited to the little killifish and can be found in many creatures across the animal kingdom including other species of fish, frogs, snails and even moles. But, unlike the other creatures, the structure of the tissue in the killifish allows them to produce fully functioning sperm and eggs from different parts of the organ, which are then brought together in a region known as the ovarian cavity. Even though all the genetic information comes from the same individual, the fish are not technically cloning

The mangrove killifish *Kryptolebias marmoratus* and the closely related *K. hermaphroditus* are the only known vertebrates that can internally self-fertilize.

It is thought that the killifish, here *Kryptolebias hermaphroditus*, evolved its selfing ability in order to survive in the highly changeable mangrove environments.

themselves as they have produced separate sex cells that then come together sexually to create a fertilized egg.

Curiously enough, the fish have never actually been observed laying eggs in the wild. Most of our knowledge of these fish and their reproduction comes from colonies kept in laboratories over the past 50 years. Some of these colonies are, extraordinarily, the direct descendants of an individual fish caught off the coast of Panama in 1994. This unique ability as a vertebrate to self-fertilize and maintain stable populations makes the fish of particular interest because, by continually having sex with itself, the fish can lose much of its genetic variation, becoming what scientists call 'homozygous'. Usually this would cause issues in the wild, as animals with little genetic variation are thought to be less able to adapt if environmental conditions change. But studies have shown that the killifish is remarkably adaptable, with individuals seemingly learning from their environment as they grow up and changing their behaviour and personalities, for example by becoming more bold, in direct response.

Even though the vast majority of the fish are self-fertilizing hermaphrodites, there are a few known male individuals. Most populations of the fish have no reported males, but in a few outlying populations in Belize almost a quarter are male. These are thought to develop either when the eggs are incubated at a lower temperature or through hermaphroditic individuals transitioning. These males can then reproduce sexually with the hermaphrodites, helping to add variation to the gene pool.

Why the mangrove killifish has evolved to reproduce in this way is thought to be linked to its irregular living quarters. Mangroves and swamps are highly dynamic environments where water levels fluctuate rapidly from hour to hour and season to season. This means that the little fish can literally find themselves left high and dry, with some reports of fish stranded up trees. As a result of being frequently trapped in ephemeral pools of water, the fish have evolved a whole suite of adaptations that include an ability to breathe air and an impressive tolerance of heat and salinity. It also explains how the fish can happily live in the dregs of a can or amongst damp leaves. It is also likely the driving force behind its selfing ability. If an individual fish finds itself trapped in a crab hole on its own, it can then just get on with populating that hole all by itself.

Duck-billed dinosaur
Bias in names

During a trek in the foothills of the Rocky Mountains in northwestern Montana, USA in 1978, local fossil hunter Marion Brandvold made the discovery of a lifetime. She found a jumble of small dinosaur bones lying within a depression in the ground. Analysis showed that these little bones belonged to around 11 individuals of a herbivorous dinosaur that were all roughly the same size, indicating that they were the same age when they died 77 million years ago. But the most striking thing about this find was the small shards of eggshell among the bones. Quite extraordinarily Brandvold had found a dinosaur nest with the baby dinosaurs still inside.

Whilst studying these fossils, palaeontologist Jack Horner (who would go on to advise for the dinosaur film *Jurassic Park* released in 1993) noticed something else. The teeth of these baby dinosaurs showed wear, indicating that they had been feeding on plant material. From this Horner deduced that they must have been fed by a parent. This was the first evidence that some dinosaurs displayed parental care, and because of this the researchers decided to name the new species *Maiasaura*, which translates as 'good mother reptile'.

The assumption, whether intentional or not, that the dinosaur providing the parental care would have been the mother, follows a long tradition of sex-bias in dinosaur names. By default, many dinosaurs have historically been named with the masculine suffix 'saurus'. This is doubtless because those studying these animals thought of them as big, brutish creatures, which they ascribed as masculine characteristics. Even to this day *Maiasaura* is one of only a handful of dinosaurs named with the feminine suffix 'saura', and presumably only because parental care is traditionally seen as a feminine attribute.

The reality is that it's incredibly difficult to determine the sex of a dinosaur. The only way to be certain is if an individual is found with eggs still preserved within it, and only if those eggs are in the lower part of the body. Otherwise, they may well have been lunch. But that doesn't mean people haven't tried. One of the first scientists to look in detail at sexing dinosaurs was the Austro-Hungarian palaeontologist Franz

Despite *Maiasaura* being named with a feminine suffix, it is currently almost impossible to determine the sex of a dinosaur.

Nopcsa. In 1905 he redescribed a sauropod fossil that he claimed was a n penis bone, or 'os penis'. While some animals do indeed have penis bones, this interpretation of what was, in reality, a clavicle bone was almost immediately met with opposition and ridicule. But that didn't stop the palaeontologist. Instead Nopcsa looked at proposed sexual dimorphism in dinosaurs, where the males and females look different due to, for example, large horns or size. In 1929 he described what he claimed were female–male pairs of ceratopsian and hadrosaur dinosaurs – those with the large frills and crests such as *Triceratops* and *Parasaurolophus*. He proposed that the differences in the skull shapes, vertebral form and even the size and shapes of limb bones between different ceratopsian and hadrosaur fossils revealed their sexes.

The fossil nest of *Maiasaura* provided the first solid evidence that some dinosaurs cared for their young by protecting and feeding them.

While Nopcsa was actually pairing up animals that lived millions of years apart and in different locations and so were probably separate species, his pioneering interpretation of these features did have a lasting impact on palaeontology. When describing new dinosaurs today, researchers now consider whether the subtle differences they observe are because the animals are in fact different species or just natural variations.

Since Nopcsa, some researchers have suggested that the shape of two little bones at the base of the tail in some dinosaurs, known as chevrons, might be used to identify males. In crocodiles these bones are used to anchor penis muscles and so are larger in the males than females. But the sizes of dinosaur chevrons didn't appear to be distributed into two neat groups. Others have looked at medullary tissue associated with egg-laying found in the bones of birds, but with equally inconclusive results.

Recently scientists renewed investigations of potential sexual dimorphism within the ceratopsians. Technological advances have allowed the detailed 3D analysis of *Protoceratops* skulls, revealing that, while sexual selection was likely driving the evolution of the animals' elaborate frills, both males and females likely sported these ostentatious ornaments, making it impossible to determine the sex of individual fossils. Rather than looking for differences in the bones of dinosaurs, it is probable that if there were any differences between the sexes they would have manifested themselves in the soft tissues that are rarely preserved.

Looking at modern birds and reptiles, many rely on bright colours and feathers to signal to members of the opposite sex, and dinosaurs probably did the same. In fact, feathers provide one rare example of sexual dimorphism preserved in the fossil record. Hundreds of specimens of the ancient bird *Confuciusornis* have been unearthed in China, allowing researchers the chance to carry out population level studies of the extinct animal. Many of the birds are so beautifully preserved it is still possible to see their tail feathers, with about 20% of the fossils showing long pendant tails. Scientists suspect that these may have been the males, while the shorter-tailed individuals were the females, although it could have been the other way around as we still have no way of determining their sex for certain.

Whether we can determine the sex of an animal that lived more than 80 million years ago may seem inconsequential, but it can still feed biases in how scientists interpret these fossils and reconstruct the animals' likely lifestyle and behaviour, such as whether the main parental caregivers were male, female, or both.

New Mexico whiptail lizard
Parthenogenesis

At first glance, whiptail lizards appear to be rather ordinary. But delve into their genetics and something extraordinary is revealed – some species are made up entirely of females. Within the whiptail genus *Aspidoscelis*, typically found in southwestern USA, all-female species have emerged again and again, to the extent that about a third of all species are now asexual, or individuals that can produce viable offspring without the input of another party. These lizards have done away with sexual reproduction and, instead, lay fertile eggs that have never encountered sperm, in a process known as parthenogenesis.

Meaning 'virgin birth', parthenogenesis is surprisingly widespread across the natural world, being found in sharks, birds, crocodiles, amphibians, snails and crustaceans to name a few. The process can be split into two forms: 'obligate' and 'facultative'. The first of these are those animals, like the whiptail lizards, which only reproduce through parthenogenesis. This has resulted in unisex species in which there are no males at all. It was only by peeking into the genetics of these lizards that scientists finally understood what was going on. Like most animals, the lizards have two copies of the chromosomes that contain the genetic information within each cell. But unlike us, these two copies are extraordinarily different. And it gets weirder. As more species of unisexual lizards were examined in detail, it emerged that some species had three copies of each chromosome within their cells. And then one species was found with four copies. It became clear that these all-female lizards were hybrids.

We're often taught that hybrid animals are sterile (think of mules and ligers) and unable to pass their genes on to the next generation, but over the past few decades scientists have gained a better understanding of the role that hybridization can play in the creation of new species. It is now known that many species, including some

The New Mexico whiptail lizard is able to produce babies without the input of any males, but that doesn't stop the females from having sex.

butterflies, fish, plants and frogs, are the result of hybridization events. It's thought that the whiptail lizards' sortie into asexuality started with a hybridization event between two sexually reproducing species. For example, the unisexual New Mexico whiptail lizard (*A. neomexicanus*) is a hybrid of the sexually reproducing little striped whiptail (*A. inornatus*) and the western whiptail (*A. tigris*). Scientists have even tested this in the laboratory, producing new species of asexual whiptail lizards that do not exist in the wild by bringing together two sexually reproducing species. The mixing of two species to create a new one explains why the chromosomes of the all-females are so radically different from each other, with one set coming from each separate species. This in turn helps to explain how the asexual reptiles maintain enough genetic diversity to persist over long periods of time. It also provides a possible explanation for how some end up with more than two sets of genes. Researchers suspect that this occurs when the all-female lizards mate with a third species of sexually reproducing lizard to produce yet another new species!

Becoming asexual is not the only queer behaviour that these lizards display, however. Whiptails do not need to have sex to reproduce, but it doesn't mean the whiptails have done away with it altogether. The all-female reptiles still engage

The western whiptail (above), found throughout the southwestern United States and down into northern Mexico, has hybridized with other lizards.

The hybrid offspring of the little striped whiptails (above) and closely related species are the fully parthenogenic New Mexico whiptails.

in sex with each other, the act itself stimulating the receiving females to produce more eggs. This behaviour is frequently referred to in the scientific literature as 'pseudocopulation', implying that sex that does not involve penetration is somehow not genuine. It is language like this that can help reinforce biases against homosexual behaviours observed in the natural world. And yet, as incredible as the evolution of an all-female species is, the whiptails are not alone. At least 80 species of reptiles, amphibians and fish from all over the planet have evolved to become unisexual.

Facultative parthenogenesis, when a species can reproduce both sexually and asexually, is more common. One of the largest species to do this is the Komodo dragon (*Varanus komodoensis*), which lives on a series of small tropical islands scattered along the southern tail of the Indonesian archipelago. Growing up to 3 metres (10 feet) long, they reproduce sexually most of the time, laying around 30 eggs which are then buried in soft soil. But on 20 December 2006, a captive female Komodo dragon called Flora took her keepers by surprise. Despite never having been in contact with a male, she laid a clutch of 11 eggs of which seven hatched and produced healthy babies. After genetic tests showed that Flora had spontaneously laid viable unfertilized eggs, scientists confirmed for the first time that Komodo dragons can reproduce via parthenogenesis. Flora's unexpected behaviour has since been found in other captive populations of dragons. Interestingly, all the females that have bred this way give birth exclusively to males.

This is to do with how the dragon's genetics work. In humans we generally talk about sex being determined by X and Y chromosomes whereas Komodo dragons (and some other species of reptiles) have Z and W sex chromosomes, in which most females are ZW and most males ZZ. When it comes to the production of sex cells, these are created via a process known as 'meiosis'. During this, the cells duplicate the chromosomes and then divide four times. This ensures that each of the four resulting sex cells only have half the number of chromosomes, so that the full number is restored during fertilization. Crucially, the chromosomes trade bits of DNA with each other during this process of division, which means that these four sex cells are not genetically identical. This means that while parthenogenesis is often referred to as 'cloning', this isn't strictly true.

For the Komodo dragons, the females are basically taking individual sex cells that have half the number of correct chromosomes and then doubling them again to make an egg. But, because the females are genetically ZW, this means that half of these resulting embryos are WW and half ZZ. Those with two Ws do not survive, but those with two Zs go on to hatch as males (as seen with the captive dragons). Komodo dragons are thought to have this ability due to their tropical lifestyle. If a female lizard is washed off a beach and ends up on an uninhabited island then it can still start a new population of dragons. And the genetics of their sex determination system means that the female will always produce a mate, even if it is her own offspring. Going back to the whiptails, the reason they are all female is because (in a very basic sense) the lizards are instead doubling their ZW chromosomes before they form the sex cells, resulting in eggs that already have a full set of chromosomes and all hatch as female.

Komodo dragons are one of the largest species to be able to reproduce via parthenogenesis, with females able to produce healthy offspring on their own.

Other champions of parthenogenesis are the insects. Not only do some species flip-flop between sexual and asexual reproduction thoughout their lives, as with the bird cherry-oat aphid and mayflies, some have also actually built parthenogenesis into their sex-determination system. In ants, wasps and bees, fertilized eggs contain two sets of chromosomes and develop into females, while unfertilized eggs with only one set of chromosomes develop into males. As the act of allowing a sperm to fertilize an egg is under voluntary control by the queens, the egg-laying females can determine the sex of their offspring by deciding whether to reproduce sexually or through parthenogenesis.

Morpho butterfly
Divided down the middle

With a dazzling flash of electric blue, morpho butterflies flit and flutter among the humid forests and along the waterways of South and Central America. Their shimmering metallic shades of blues and greens are not the result of pigmentation, but an example of what is termed 'structural' colour. The colour is produced by the precise shapes and angles of the scales that coat their wings and reflect the light back in just such a way as to produce the gloriously vivid colours that have beguiled scientists and collectors alike.

But some individuals look strikingly different. With a line running down their body so perfect it looks almost artificial, the butterflies have wings that look entirely different to each other. This is because one pair of wings are genetically male and the other genetically female. Known as gynandromorphs, these animals have both male and female tissue within the body of a single individual. This is distinct from hermaphrodites, which are able to produce both male and female sex cells at some point in their lifetime. Gynandromorphs have been found in a huge range of different animals, from lobsters, spiders and bees, to lizards, snakes and occasionally mammals. But the most distinctive and eye-catching gynandromorphs are often found in butterflies and birds, particularly in species that are highly sexually dimorphic (when the males and females look dramatically different from one another).

While the term 'gynandromorph' only really started to appear in natural history writings around the 1870s, the spectacular appearance of many gynandromorphs was almost certainly noticed by natural historians and scientists long before that. It is difficult to know when they were first observed, but it is likely that they were historically simply referred to as intersex, dual-sex and, in some cases, 'perfect' or 'true' hermaphrodites.

It has been suggested that perhaps 1 in every 10,000 butterflies is a gynandromorph, although in reality it is almost impossible to know for certain. The subtle differences in wing shape, size and colour reveal that these morpho butterflies are gynandromorphs, with one half of the insect male and the other half female.

Gynandromorphs are often reported from highly sexually dimorphic species such as butterflies and birds. In this red cardinal, the male is on the bird's left, the female on the right.

There is an example of this in a 1761 work by Jacob Christian Schäffer, who described a bilateral gynandromorph spongy moth (*Lymantria dispar*) in *Der wunderbare und vieleicht in der Natur noch nie erschienene Eulenzwitter* (roughly translated as: 'The wonderful hermaphrodite moth, which perhaps has never appeared in nature before').

How common these animals are is tricky to figure out. Different estimates suggest that less than 1% of crustaceans are dual-sex, that roughly one in 17,000 spiders are gynandromorphs and that one in around 10,000 butterflies are. In reality, it is impossible to know for certain because many gynandromorphs go about their business entirely unnoticed; saying 'gynandromorph species' implies that all members of the species are gynandromorph. There are a lot of reports of butterflies and birds, groups studied in greater numbers, but relatively few of centipedes and frogs. The perceived rarity of gynandromorphs may also be down to the different types there are, with some harder to spot than others.

Gynandromorphs come in at least three different flavours, with a number of other hypothesized complexions. The first of these are 'polar' or 'anterior/posterior' gynandromorphs. In these individuals the front half is one sex and the back half the other, and is evident in a number of species, including wasps, mosquitoes and fairy shrimps. In some species these polar gynandromorphs occur at a higher rate than would be expected for random mutations alone, suggesting other causes, such as temperature fluctuations during embryogenesis and the potential that it might be a heritable characteristic, passed from parent to offspring. The second type of gynandromorphs are 'mosaics'. In these individuals, the male and female cells are scattered seemingly randomly throughout the body of the individual, resulting in, for example, a butterfly with a patch of male cells in an otherwise female wing. The most well-known, though, are the 'bilateral' gynandromorphs. This is when, as with the morpho butterfly example, an individual is divided perfectly down the middle, with male on one side and female on the other.

But it has been suggested there are potentially a number of other types of gynandromorphs out there yet to be uncovered. For example, it has been hypothesized that 'spiral' gynandromorphs might exist with individuals split into quarters, each opposite corner being of the same sex. Whether or not these individuals exist is exceedingly difficult to tell. It could be that there is only a subtle mix between male and female, which is difficult to identify, leading us to underestimate the number of gynandromorphs in general. Or it could be that the developmental pathway that leads to these individuals is more constrained than we know.

This leads on to how exactly do gynandromorphs form, and the short answer is, well, no one is really sure. A number of different ideas have been floated, and the actual answer almost certainly depends on which species is looked at combined with what type of gynandromorph is produced. This should probably come as no surprise considering the vast array of animals in which split-sex individuals are found. The way in which sex is determined in a chicken is vastly different to how it comes about in a wasp, which in turn is not the same as a butterfly.

What is certain is that gynandromorph individuals are teaching scientists that the processes behind sex determination are nowhere near clearcut. A complex dance is at play between chromosomes, hormones, proteins and chance. The more we learn about sex, the more we find exceptions to the rules.

Western lowland gorilla
Queer behaviour in apes

When it comes to understanding homosexual behaviour in humans, many scientists have turned to apes, given our evolutionary closeness. Within the lowland (*Gorilla gorilla*) and mountain (*Gorilla beringei*) gorilla species found across the central tropical belt of Africa, scientists have made innumerable observations of queer behaviours. Both species follow the same basic social structure, involving family groups, called troops, that are usually made up of one silverback male and multiple adult females and their various offspring.

As young females come of age, they leave their family group to find a new troop of unrelated adults. As young males reach sexual maturity they are chased out of their troop and either take up a solitary life in the forest or band together with other exiled males. Known as bachelor groups, the males within these cohorts form a complex network of homosexual pairings. Typically, the older individuals living within these all-male collectives have favourite blackbacks with whom they have sex, which can be initiated by either party. Some individuals are effectively monogamous, while some males have multiple partners. The older gorillas can become so fond of favourite partners that they guard and aggressively defend them from other individuals seeking to mate with them. Things can even turn violent when a new male joins the group – older males aggressively display, chase and physically attack each other in order to win the affection of these newcomers.

It is not just the males that frequently have sex with each other though. Within the family troops, the female gorillas regularly engage in homosexual mating that often involves rubbing their genitals together while grunting, whimpering and growling in pleasure. What is striking with both male and female homosexual activity is that it appears to be more intimate than heterosexual sex. In both sexes, the queer activity frequently takes place face to face, in comparison to the more usual

Some young male western lowland gorillas form bachelor groups within which a complex network of homosexual relationships can form.

As young female gorillas grow older they typically move into new troops, where they form sexual and affectionate relationships with both the dominant silverback and other females.

back to face sex observed between males and females. There also seems to be more grooming and sexual contact amongst the queer pairings, with one study that looked at lesbian behaviours finding that, on average, lesbian sex lasted five times longer than heterosexual sex.

All species of the larger apes (which includes orangutans, gorillas, chimpanzees, bonobos and humans) and many species of gibbon show some degree of queer behaviour, often in the form of homosexual and bisexual interactions, with some fascinating expressions. Both male and female orangutans have, for example, been observed masturbating members of the same sex. Male orangutans have been seen giving each other oral sex, while a female was once observed manually stimulating her partner with her fingers whilst simultaneously stimulating herself with her own foot. Talk about flexible behaviour.

One particularly notable case is that of Donna the chimpanzee. Donna was born at what is now the Emory National Primate Research Centre in Atlanta, Georgia, USA. As she aged, researchers noticed something different about her. From a very young age she played more with the larger, dominant males than the other young female chimps, and went on to develop a stature, face and fur more typical of males. As an older individual, she competed with the males in the dominance hierarchy, all of whom appeared to accept her as one of them. Primatologist Dr Frans de Waal, who watched Donna grow up at the primate centre, later described her as 'a largely asexual gender-nonconforming individual'. Whether or not animals can have a gender (and what this would actually mean in reference to a chimpanzee) is difficult to discern, but Donna clearly did not fit the typical binary we expect of chimpanzees.

Conversations about queer behaviour in apes cannot ignore one glaring example – the bonobo (*Pan paniscus*). Long thought to be a smaller subspecies of the chimpanzee, the bonobo was recognized as its own distinct species in 1928. Separated from each other in evolutionary terms by millions of years and the Congo River, the bonobo and chimpanzee societies couldn't be more different from one another. While chimpanzees typically settle their differences through fighting, bonobos do so largely through sex. A lot of sex.

Bonobos display potentially the largest repertoire of homosexual behaviour of any animal. Males and females, from adolescents to adults, engage in oral sex, masturbation, sex and even open-mouthed kissing with tongues, rarely observed in species outside of humans. Males even hang from trees with their arms while fencing with their erect penises. But it is the intense female relationships that are of particular note. Bonobo society is matriarchal, meaning that it is dominated by the females who form especially strong social bonds. These bonds are regularly reinforced with generous doses of sex. One of these behaviours is known as 'genital-genital' – or GG rubbing. When engaging in GG rubbing, the females lie face to face with their legs spread to maximize contact of their large clitorises. Sometimes one even wraps its legs around the other so that they are carried off the ground. Throughout this sex, both females rhythmically swing their pelvises from side to side, all the while grinning widely and vocalizing as they reach orgasm. So intense is the GG rubbing, and so distinct the bonobo clitoris in its size and positioning, some scientists suspect that this lesbian activity has actually been driving the evolution of the bonobo's genital architecture.

Domestic sheep
Can animals be gay?

When talking about the queerness of animals, scientists frequently refer to an animal displaying homosexual 'behaviour' or gay 'activity'. This is because it is impossible to ask an animal about their sexuality and difficult to follow a single individual its entire life to observe its mating habits, making it exceedingly hard to know if an animal is truly gay or not. If we were to use human definitions for animals, then most would not be classified as gay or lesbian but something more akin to bisexual, but even this becomes problematic when trying to assign sexuality onto an individual animal. The domestic sheep (*Ovis aries*), however, offers us a unique opportunity.

Sheep were domesticated some 10,000 years ago in the Middle East, so people have been living alongside them, intimately observing their behaviours, for an incredibly long time. This, coupled with their sheer numbers (estimated to be in the region of 1 billion animals), means that humans have had more opportunity to observe them than many other living creatures. Farmers take a keen interest in the sex lives of their livestock and have long noted that some rams are simply uninterested in ewes. This is sometimes put down to a 'low libido', with vets even offering injections to try to remedy it. The more straightforward explanation is that they are simply queer.

Studies have found that roughly 8% of all rams show a consistent preference for other rams, even when given the opportunity to mate with ewes in heat. This suggests that sheep may be the first known animal, apart from humans, in which genuinely homosexual individuals have been identified. Some of these experiments have, however, been met with controversy. Because rams need to do their job and breed, the farming industry has spent time and money trying to identify gay tendencies and treat them. Experiments have looked at biological signatures in blood, such as hormone levels, and studied the brain structure of gay sheep. Nothing suggests that any of this would apply to humans, but this creeps uncomfortably close to the

Big horn sheep have been described as living in 'homosexual societies'
due to the amount of queer sex that takes place amongst the rams.

The domestic sheep is the only animal, apart from humans, in which we can be relatively certain that some individuals are genuinely gay.

medicalization of queer folk, the potential for a 'cure' and even eugenics. Studying homosexual animals can have political and societal consequences of which those doing the research need to be acutely aware.

Intriguingly, the gay behaviour of sheep is not limited to domesticated animals. Research has found that their wilder cousins are also queer. Bighorn sheep (*Ovis canadensis*) are found in the mountains throughout central and western North America. Named for the males' large, curving horns, the sheep form sizeable herds in which the males fight for access to the females. The sheep have, however, been described as living in 'homosexual societies'. The rams in these herds regularly engage in homosexual courtships and sexual activity with, typically, the older males initiating the behaviour with younger rams. The rams bow their heads, rub their horns against each other's bodies, and taste their partner's urine in a behaviour known as 'flehmen' during which males bare their teeth; it is usually thought to be a way for males to know if a female is on heat. This is often done while the ram is fully erect, before it mounts and penetrates the younger animal. In the closely related Dall sheep (*Ovis dalli*), males also lick each other's genitals.

In fact, queer sexual activity is so pervasive among male bighorn sheep that it can sometimes be difficult for females to get the attention of the dominant rams. Some females on heat have been observed mimicking the behaviour of young males to encourage the rams to mount and mate, which they do, ironically, only when they think the females are males. In a further twist, it has been reported that some males mimic the behaviour of females to avoid participating in the queer behaviour of the other rams. Whether, like the domestic ones, some of these rams are what we would consider gay, with a continual preference for other males, is still tricky to answer, but the parallels between the wild and domestic species are notable nonetheless.

In recent decades, our understanding of queer animals has also expanded as our knowledge of the natural world has deepened. Japanese macaques (*Macaca fuscata*), a species of monkey living from the wintery mountains to the subtropical forests in the Japanese archipelago, have also consistently indulged in queer activity. They typically form groups, known as troops, consisting of roughly 30 individuals, although occasionally up to 100 depending on the availability of food. These troops are organized around different matrilines, in which the power and dominance is passed from mother to daughter. Studies over the past four decades have found that some female macaques show a consistent preference for mating with other females, even when there are sexually active males on the scene. The researchers found that females court each other before mounting in up to eight different postures, in a kind of monkey kama-sutra, whilst heavily thrusting and gazing intently into each other's eyes. The scientists concluded that these females were having sex with each other purely for pleasure and, as they also mated with males, that they were seemingly genuinely bisexual in nature.

Now this does not mean that sheep and macaques are the only animals that could be considered to have a sexuality that we would define as homosexual or bisexual. It is highly likely – if not almost certain – that there are plenty of other animals out there where individuals have a consistent preference for members of their own sex, it is just that sheep and primates are both intensely studied animal groups. The more we look, the gayer the world undoubtedly is.

Saharan cypress
Androgenesis

While exploring the depths of the Sahara Desert in the 1850s, the British clergyman and ornithologist Henry Baker Tristram started a curious rumour that there were conifer trees growing in the parched, sun-baked dunes of the desert. As extraordinary as this sounded to a naturalist more accustomed to seeing conifers dripping rain or dusted in the snow of northern forests, the carved wooden saddles of the local Tuareg people suggested that it was more than just hearsay. But it would take another 60 years for Europeans to discover what the Tuareg had known for millennia. It was the French commander Captain Maurice Duprez who finally set eyes on the Saharan cypress (*Cupressus dupreziana*) and sent specimens back to Europe to be scientifically described in 1924.

Far from a stunted, gnarled tree clinging to life, the Saharan cypress strikes a remarkable silhouette amongst the dunes and rocky outcrops. Its distinctive blue-hued crown billows from a thick trunk, and robust branches jut out from the base and curve upwards. Its roots extend outward, snaking across the rocky substrate on the hunt for the smallest drops of water. With a reddish-brown bark streaked by long deep fissures, the largest known tree reaches 12 metres (39 feet) in girth and 22 metres (72 feet) in height and is thought to be at least 2,200 years old. In the early twentieth century naturalists thought the trees were doomed to extinction. Duprez only found a handful of them growing in the southeastern corner of Algeria, and the widespread belief was that the seeds were all infertile. It turned out, however, that the low fecundity of the seeds hinted at the truly unusual nature of these trees: they only pass their male genes on to the next generation. When it comes to sex in plants, all bets are off.

Most conifer trees are termed 'monoecious'. This means that the plants produce both 'female' flowers that contain the ovule, and 'male' flowers that produce the pollen.

The Saharan cypress is an extraordinary tree as not only is it a conifer that grows in the desert but it is the only species of plant known to reproduce via androgenesis.

This is roughly, but not exactly, analogous to hermaphrodites in animals, which produce both male and female sex cells. While most plants reproduce sexually, some instead reproduce asexually through a process known as 'apomixis'. This is similar to parthenogenesis in animals (*see* New Mexico whiptail lizards), meaning that the viable seed arises from an unfertilized female sex cell and so no male contributes to the next generation.

But the Saharan cypress changed everything we thought we knew. With conifers, each pollen grain actually contains two 'male' sex cells. Typically, when the pollen meets the ovule, it sends down one of the cells to fertilize the 'female' sex cell while the other gets discarded. With the cypress it seems likely that, rather than discarding the second male sex cell, it keeps it and discards the female sex cell instead. This results in only the male genetic information being passed on. This type of reproduction, known as paternal apomixis, had never been seen before in plants.

Also called androgenesis, this form of reproduction is exceedingly rare in the natural world. While the cypress is the only plant known to do this, across the rest of nature there are a handful of other examples, including the little fire ant (*Wasmannia auropunctata*), the stick insect (*Bacillus rossius*) and the freshwater fish (*Squalius alburnoides*). Androgenesis was confirmed in the cypress through experiments in which researchers used pollen from the Saharan cypress to fertilize the closely related Mediterranean cypress (*Cupressus sempervirens*). After 15 years they could see that the offspring looked just like the Saharan cypress, rather than a mix of the two species, and genetics confirmed that they were indeed pure Saharan cypress trees and not hybrids. It has been suggested that the androgenesis is behind the low fertility rate of the seeds as only around 10% of all Saharan cypress seeds are viable. But no one is quite sure why this would be the case when the reverse – maternal apomixis – doesn't seem to have a viability problem.

A total of 233 known wild cypresses survive on the fringes of the Sahara, making it one of the most endangered trees in the world, and botanic gardens around the planet, from France to Australia, now grow them as an insurance should the worst happen. The trees face a number of threats in their little corner of the world; once felled for timber, they are now more likely to be overharvested for firewood by tourists and impacted by the climate crisis increasing the intensity of droughts and the likelihood of flash floods, which can knock over the shallow-rooted trees.

But we should not give up on on these old sentinels of the sand. While the prevailing notion is that asexual reproduction is an existential death knell for a species, reducing diversity and limiting their ability to adapt, some of the trees standing today have been gently persisting in the warm breeze of the Sahara since Hannibal traversed the Alps with his army of elephants. They have grown ever taller and ever wider throughout the reigns of kings, queens, pharaohs and emperors. The trees continue to propagate on their own, and with local communities helping to protect them, the Saharan cypress seedlings of today may grow tall for millennia to come.

The Saharan cypress is one of the world's most endangered trees, with only around 10% of its seeds viable. But no one is quite sure why this is the case.

Bicolour parrotfish
Sex-fluid fishes

If you dive on a coral reef, a quarter of all the fish species swimming around you will be a form of hermaphrodite. This means that individuals produce sperm and eggs at some point in their life. And parrotfish are no exception. These brightly coloured fish are often one of the most numerous groups of fish inhabiting reefs from the Caribbean Sea to the Pacific Ocean. Named after their beak-like mouth, the fish form an integral part of the coral reef ecosystem. After they break off and eat the coral for the algae that grows on it, the parrotfish poop out a fine sediment that goes on to form a part of the pristine white sandy beaches that many people enjoy on their tropical holidays.

Many parrotfish are highly sexually dimorphic, meaning that the males look vastly different from the females. Typically, the females have darker, more drab colours while the males splash vibrant hues of pink, green and blue. In some species the juveniles can also look radically different from the adults. And to add into that mix – most species change sex. The majority of parrotfish, such as the bicolour parrotfish (*Cetoscarus bicolor*), go from female to male when they reach a certain size, but a few species go in the opposite direction, from male to female. Only one known species does not change sex at all. This dramatic difference between the sexes and ages, and the tendency of the fish to liberally switch sexes, led scientists to describe hundreds of species of parrotfish before realizing that there are actually only around 90.

Parrotfish are what are known as sequential hermaphrodites, meaning they produce male and female sex cells, but at different points in their life. In contrast, simultaneous hermaphrodites produce both at the same time (*see* Mangrove

Male and female parrotfish look so vastly different from each other that researchers initially described many more species than actually exist. The bicolour parrotfish is no exception, with the females (bottom) changing sex to become male (top) when they reach the correct age and size.

killifish). Sequential hermaphrodites are further divided into three categories. Most parrotfish are what are known as 'protogynous', as many species start out as female and change to male when they reach a larger size. This is the most common form of hermaphroditism in fish, and parrotfish may do this because of their mating system in which a single large male has a harem of smaller females. It has been argued that under these circumstances there is little advantage for a fish to start life as a smaller male as it would be unlikely to defeat the larger male to mate. So the fish can increase their reproductive success by first developing as females before switching sex later in life when larger in size.

Some species go in the other direction and change sex from male to female, known as 'protandry'. Clownfish, a group of around 30 species of fish that live in and amongst sea anemones, are one of the most well-known examples. Typically the fish, such as the ocellaris clownfish (*Amphiprion ocellaris*), live in rigorously hierarchical

The ocellaris clownfish lives in rigorously hierarchical societies and so when the dominant female dies the top ranking male changes sex to take her place.

groups in which a larger, dominant female and male pair reproduce, with a number of smaller males in the group. If the female dies, then the largest male simply changes sex to take her place and mates with the next largest male, with everyone shifting up in the pecking order. If the film *Finding Nemo* was strictly accurate, when Nemo's mother died at the start of the film his father should have changed sex to replace her and then start having babies with the protagonist.

Other species that can go either way are called 'bidirectional' hermaphrodites, such as the blue-banded goby (*Lythrypnus dalli*), whose sex is dictated by social status. Regardless of what sex the fish first starts out as, the individuals either change or stay as male or female depending on where they eventually fit in the social hierarchy of the group. Those toward the lower end of the dominance hierarchy become female, while those at the higher end become or are males. Simultaneous hermaphrodite individuals produce male and female sex cells at the same time. This includes species such as the tripod fish (*Bathypterois grallator*), which has been found down to 4,720 metres (5,500 feet) in the ocean depths. Because of their extremely remote environment, the fish have to be certain that when they meet an individual of the same species they can have sex and reproduce, so both individuals release their sperm and eggs into the water where they mix and fertilize.

While we've limited ourselves here to the sex-fluid species of colourful coral reefs, they are far from alone in the natural world when it comes to switching up their sex. Around 65,000 species of animals are known to be hermaphroditic, ranging from the dusky ancylid limpet (*Laevapex fuscus*) and the cottony cushion scale insect (*Icerya purchas*) to the intertidal isopod (*Gnorimosphaeroma oregonense*) and the Asian swamp eel (*Monopterus albus*).

But the true champions of hermaphroditism are plants. It is thought that around 85% of all flowering plants – equal to roughly 212,500 species – are hermaphroditic, as the flowers contain both male and female structures. Other plants, such as conifers, are what is known as 'monoecious' as individual trees produce separate male and female flowers (*see* Saharan cypress), while those termed 'trioecious' have male, female and hermaphroditic flowers on individual plants. Truly, when it comes to switching up sex, nature goes in all sort of directions.

Swans
Male couples as parents

On a brisk spring day in 1923, John Peacock Ritchie was rambling around the edge of a loch in Renfrewshire, to the west of Glasgow, Scotland, when he came across a pair of mute swans (*Cygnus olor*) sitting on a nest. The ornithologist and collector stopped to snap a photograph of the couple before, as was common for the time, chasing the birds off their nest of twigs and reeds to see its contents. To his surprise, the nest was empty. A week later, Ritchie returned to the nesting pair and still there were no eggs. At first, he suspected that railway workmen nearby had stolen the swans' eggs, but on a third visit to the pair the nest remained bare.

It was only after revisiting the nesting swans a few more times over the following weeks and analyzing the photograph he had taken, that Richie finally realized what was going on. The large black knobs at the base of the swans' bills revealed that the two birds were in fact both males. The picture that Ritchie took is now thought to be the first confirmed photograph of queer animals. The editors of the *Scottish Naturalist,* the publication in which Ritchie reported the queer swans, tried to explain this behaviour based upon historic documentation of female birds presenting as males (*see* Common pheasants), thus implying that what Ritchie saw was actually a heterosexual couple. But Ritchie himself referred to an earlier record in 1885 of two female mute swans that 'had lived the same way', nesting and laying eggs together. Since these initial early records, it is now known that queer behaviour is fairly common amongst swans, and in particular in the black swan (*Cygnus atratus*).

Black swans are one of two species of swan endemic to the southern hemisphere. Commonly found serenely gliding around the wetlands and coastal islands of their native Australia, black swans have been introduced to New Zealand, while

Up to 20% of all pairs of black swans form queer couples, with these homosexual pairings often having larger territories and being more successful in rearing chicks.

Published in the *Scottish Naturalist* in 1926, this photograph of two male mute swans, taken by John Peacock Ritchie, is thought to be the first confirmed photograph of queer animals.

feral populations can also be found in various European countries, the USA, Japan and China. As is typical for most swan species, the birds generally form socially monogamous pairings. While swan couples may court, nest and raise chicks together, they are not necessarily sexually monogamous. Almost 40% of black swan broods contain at least one chick that is the result of 'extra-pair paternity', a fancy way of saying that the chicks are not related to the male raising them. While most of these pairings are between males and females, a significant number are queer. In any one year up to 20% of all black swan couples are between two males, while occasionally swans form something of a 'throuple', with two males and a female displaying to each other and looking after the nest and chicks.

All black swans court each other with a 'wave-like' ducking activity, in which the birds hold their heads and necks parallel to the water's surface, before dipping their head into the water and thrusting forward at the same time. They repeat this awkward performance a number of times to cement the bond. It is at this point that the queer partnerships really hold their own – homosexual pairs and trios of black swans can often maintain much larger territories than their heterosexual counterparts. This might be intuitive for the groups of three, but it is thought that the gay swans are better able to pool their strengths and therefore more effectively chase off any rivals.

After building a nest together, the black swans have several choices depending on their situation. The trios can breed and lay eggs, with the two males and a female caring for the brood and raising the chicks together. Female pairings are also able to lay eggs, with one of the females soliciting nearby males to mate before returning to their same-sex partner to lay. Gay swan pairings have to do things differently. Sometimes one of the males mates with a female and then claims the resulting eggs and rears the brood with its male partner. Another tactic involves the male pair chasing heterosexual swans off their nest and then raising the clutch of eggs as their own.

Scientific literature often describes queer animal behaviour using terms such as 'maladaptive', implying that it is somehow – evolutionarily speaking at least – detrimental to a species. The ubiquity of homosexual behaviour in animals, from beetles to bats, challenges that conclusion. These statements are often based on the notion that the only benefit of sex is reproduction, which we know is not true. But the queer behaviour of black swans also contradicts the assumption that homosexuality is maladaptive, because male pairs are often more successful at raising chicks than their heterosexual counterparts with, on average, around 80% successful parenting efforts by homosexual couples compared to 30% by heterosexual couples.

Queer swans' success at raising chicks may be attributable to maintaining larger territories, but it could also be that female black swans expend greater energy in incubation efforts, which limits future successful reproduction in male–female pairs. Homosexual swans are frequently more equitable when it comes to parental duties. Rather than being pigeonholed as a curious side note in animal behaviour, such as Ritchie's observations in 1923, queer individuals can help us question and challenge long held assumptions in evolutionary biology.

Green sea turtle
Temperature-dependent sex determination

At school we're taught that what an animal looks like – known as its phenotype – is determined by its genetics and not its environment. But, as with many aspects of biology, things are never quite as simple as they seem. For example, the sex of many species of reptiles and fish is determined by how genes in each of their cells respond to the temperature at which the eggs are incubated. This is known as temperature-dependent sex determination. The two types of this form of sex determination are known rather unimaginatively as 'Pattern I' and 'Pattern II'.

The green sea turtle (*Chelonia mydas*), a marine reptile found swimming in tropical waters around the globe, nests on sandy beaches from Ecuador and Florida to Yemen and Australia. Like other sea turtles, they exhibit the Pattern I form of temperature-dependent sex determination. Within this pattern, eggs incubating below a certain temperature will hatch as one sex and above it, the other. With the sea turtles this magic number is usually around 29°C (84°F), with lower temperatures favouring males and higher temperatures females. As the temperature of the sand in which the turtles nest is usually warmer toward the surface, the eggs nearer the top will hatch as females while those near the bottom of the clutch (and thus cooler) will hatch as males. The opposite sequence can also occur. The tuatara (*Sphenodon punctatus*) is a curious reptile from New Zealand that looks like a lizard but actually belongs to a completely separate group (order Rhynchocephalia). In these animals, it is the lower temperatures that produce mostly females, while a higher one produces males.

With Pattern II things are a little more complicated. In these animals, such as saltwater crocodiles (*Crocodylus porosus*), there isn't a single temperature point, but two. Any eggs incubating below 30°C (86°F) hatch as female, those between 30–32°C (86–90°F) are more likely become male, while any over 32°C (90°F) will

The sex of green sea turtles is determined by the temperature at which the eggs are incubated, meaning that climate change could affect the reptile's sex ratio.

The climate crisis is also affecting central bearded dragons, with warmer temperatures causing genetically 'male' lizards to develop as female.

also be female. Within some Pattern II species, eggs developing at the threshold temperature will occasionally even develop as intersex.

How temperature-dependent sex determination actually works is thought to be down to the levels of certain hormones. While an organism's genes typically affect which sex hormones are produced in chromosomal sex determination, the environment provides the trigger for crocodiles, turtles and some species of fish and lizards. As a result, these animals have done away with sex chromosomes altogether. For these species, the temperature of incubation dictates the activation of various hormones, such as aromatase, which in turn influence which suite of hormones encoded in the genes are turned on or off.

While we still don't fully understand the ins and outs of this process, it is clear that this method of sex determination has worked for crocodiles and turtles for hundreds of millions of years. But the past 150 years of human activity could be a challenge too far. As the world has rapidly warmed due to the burning of fossil fuels, scientists have been studying the impact on the sex ratio of crocodiles and their relatives. One study found that an increase in average temperatures of between just 1.1°C and 1.4°C (2 and 2.5°F) in American alligator (*Alligator mississippiensis*) nests would skew the sex ratio toward 95% male hatchlings, while any additional heating will likely swing things toward more females. If these animals cannot move to cooler northern climates, it could seriously hinder their ability to survive.

There are other, slightly more complex ways in which changes in temperature might affect the sex ratios of reptiles. Under certain conditions, genetically male individuals of some species of lizards can develop as females. This can be seen in bearded dragons, a group of mostly mid-sized lizards native to the arid regions of Australia that get their name from the series of spikey scales on their chins which they can inflate. Whereas in humans we generally talk about sex being determined by the X and Y chromosomes, in some lizards these are instead Z and W chromosomes, with individuals with ZZ chromosomes expected to be male and those with ZW female. But with central bearded dragons (*Pogona vitticeps*), embryos with ZZ chromosomes incubated at temperatures above 32°C (90°F) develop as female. These 'male' chromosome individuals have female genitalia and can lay eggs. This suggests that at least some of the characteristics considered to be 'male' are determined by external factors, similar to how temperature interacts with genetics, even in species that have sex chromosomes. Being adapted to the arid parts of the country means that bearded dragons may also be impacted by the climate crisis. As average temperatures across Australia continue to increase, one of the possible consequences is that some species of bearded dragons, such as the central bearded dragon, may become all female populations (*see* New Mexico whiptail lizard). And the impact of global warming is even more profound in species with temperature-dependent sex determination like crocodiles and sea turtles.

This research highlights the potential unforeseen impact that the climate crisis might exert on species, and the complex interactions that occur between genetics, phenotype and environment. It also shows how there is no one route to determining the sex of a species, and how flexible this system can be. Think about those green sea turtles that are either male or female solely as a result of where the egg they came from was buried in the nest.

Giraffe
Homosexuality in the mainstream

In 1956, Anne Innis Dagg made history. At the age of 23, she became the first person we know of to study wild animals in Africa scientifically, and only the second in the world to start a long-term study of a single mammalian species. Innis Dagg made the long and arduous trip from her home in Ontario, Canada, all the way to the edges of Kruger National Park, South Africa, to fulfil her dream of studying giraffes (*Giraffa camelopardalis*) in the wild. Battling the entrenched sexism of the scientific world, she became a true trailblazer. She also became the first to document the extensive queer behaviours of wild giraffes. Despite pushback from the scientific establishment, Innis Dagg continued to publish on this subject, specifically citing her lesbian students as inspiration to tell the world what she was seeing in nature.

In 1984 Innis Dagg also published the first analysis of homosexual behaviours in mammals. She studied the records for 125 mammalian species and concluded that 'adult homosexual behaviour is widespread in male and female mammals'. It is difficult to understate the significance of this paper. Historically, homosexual behaviour in animals was seen as a curious oddity (*see* Swans). It was largely viewed as a peculiar expression of dominance or frustration on the fringes of normal behaviour, and certainly not something happening with sufficient frequency to warrant being taken seriously.

But Innis Dagg's giraffe observations turned that on its head. In some populations of giraffes over 94% of all sexual activity observed was homosexual. Typically this was between two males, but not always. Giraffes are perhaps best known for their fashionably patterned coats and elegantly prolonged necks. Primarily, these necks allow them to reach the tops of trees and snack on leaves inaccessible to other animals milling about below. But male giraffes also use their necks for a unique courtship display. Far

Male giraffes engage in a range of gay behaviours, including entwining their necks, sniffing each other's genitals and eventually mounting and having sex.

The high levels of gay behaviour observed in giraffes could be related to the fact that the males live predominantly in bachelor groups for most of their lives.

from being a solid beam, their necks are actually quite flexible, allowing the animals to wrap them around each other in a behaviour known as 'necking'. Males frequently engage each other with necking, which tends to start with the giraffes facing each other, and then gently rubbing their necks on the other's body, head and thighs, before entwining them. This is often accompanied by a little sniffing and licking of genitals, and usually results in an erection. Occasionally, the giraffes also display the 'flehmen' response (baring their upper teeth to detect the pheromones in urine) associated with sexual arousal between males and females. Finally, males mount each other with erect penises and ejaculate. This behaviour isn't limited to pairs of males, as sometimes bigger groups of males gather to neck and mount each other over and over. Females also irregularly mount each other, but typically they don't engage in the necking ritual.

With giraffes living in sex-segregated societies, homosexual behaviour is astonishingly common and not that surprising given that, for most of the year, the males spend most of their time hanging around with each other. But some have suggested that this is just dominance behaviour, despite the clear sexual arousal of the individuals involved. This highlights the frequent double standards to which homosexual behaviour in animals is held. If a male giraffe necks with a female, licks her genitals, and gets an erection before mounting and ejaculating, it is unequivocally described as a sexual behaviour. Yet if two males do the exact same thing, it is no longer deemed sexual but is called dominance behaviour instead.

In 2019 this debate even managed to cause something of a split in the UK Labour party and hit the mainstream media. At an awards event for the LGBTQ+ news website PinkNews, the Labour MP and LGBTQ+ ally Dawn Butler said: 'They talk about teaching people or children to be gay. They don't want people to be taught to be gay. I want to know this, right: if you can teach 'gayness' – if that's even a word – if you can teach gayness, who speaks giraffe? Because 90 per cent of giraffes are gay. So, if you can teach it, I want to know who the hell speaks giraffe?' This sparked a response from Lachlan Stuart, the then adviser to the Labour leader, who called Butler's remarks 'a ludicrous, offensive, homophobic claim' and stated that giraffes do not display 'gay behaviour' because, 'There's no romance. No courtship. No affection. No pair bonding'. He insisted that it was all dominance behaviour, before adding, rather perplexingly, that giraffes were his favourite animal. Unsurprisingly, the UK media had a field day with this mild ruckus. 'Gay giraffe row splits upper ranks of Labour party'.

While these headlines may have been blowing things slightly out of proportion, in addition to mixing up sex with gender, it is mostly impossible to know for certain if an animal is actually gay (*see* Domestic sheep). But, as we have seen, the vast majority of sexual behaviour amongst giraffes is very queer indeed. And, due to the social structure of giraffes, with only a tiny fraction of individuals ever getting the chance to breed, it is highly possible that, when it comes to sex, some males do only ever engage in gay behaviour.

Common ash
Sexual spectrum

We often describe plants as having structures that are either 'female' or 'male'. While not perfect, this is the common terminology used when we are talking about organisms which are so vastly different and separated by hundreds of millions of years of evolution. This is largely based on the idea that 'female' plants produce larger sex cells from which the embryo grows. But sex is more complicated than this as it relies on a complex interaction between genes, hormones, environment and chance. And with plants, there are additional layers upon layers. For example, the common ash (*Fraxinus excelsior*), also known as the European ash, is a flowering plant in the olive family that was historically found growing from western Europe all the way east to the Caucasus (in recent times, the species has been severely hit by the ash dieback disease). Ash trees not only have individual male and female trees, but also trees that exhibit a spectrum between the two.

In general, flowering plants reproduce sexually by three main methods. Plants that have historically been referred to as 'hermaphrodite', 'bisexual', or 'perfect' flowers, and are increasingly being called 'cosexual', use the first of these methods. Their flowers contain both 'female' and 'male' structures, meaning that they produce both pollen and ovules from the same bloom. The majority of flowering plants are cosexual and while in theory they can self-fertilize, in reality it varies from species to species and how they have organized the structures within each flower. Usually they avoid 'selfing' in order to maintain genetic diversity which, in turn, helps protect against disease and environmental change (*see* Chinese shell ginger).

In the second method, plants keep the male and female structures entirely separate. Those plants with separate male and female flowers on the same individual plant are known as 'monoecious', from the ancient Greek for 'one house'. In these plants, such

The common ash tree is unusual in that it produces not only 'male' and 'female' flowers but also cosexual flowers which themselves have male and female structures.

Ash trees produce different combinations of sex along a spectrum, from only producing 'male' flowers at one end to only producing 'female' flowers at the other.

as the mayapples (*Podophyllum*) found from Afghanistan to China, Canada and the USA, some flowers contain only the stamen and others only the pistil. In a similar way to cosexual plants, these plants often have methods to prevent selfing, usually by having different flowers open at different times of the day or year.

Finally, the third method of reproduction is known as 'dioecious', meaning 'two houses'. These plants have separate 'male' and 'female' flowers found on separate plants. This is more akin to how many people assume animals organize things, but as with animals, things are never quite as simple as they first appear (*see* European yew).

Plant reproduction just doesn't fit into neat little boxes. Plenty of flowering plants don't adhere to any of these systems. And this is where the common ash comes in. The ash trees not only have individual male and female trees, but also trees inbetween the two. They are 'trioecious', you guessed it, from the Greek for 'three houses'. These plants produce trees that have male, female or cosexual flowers. This is an exceedingly rare mating system in the natural world – only around 0.03% of known flowering plants show this organization of the sexes. But trioecy is not just limited to plants.

The bisexual mussel (*Semimytilus algosus*) was initially thought to be a hermaphroditic species (*see* Bicolour parrotfish), but was later identified as the first known species of trioecious mollusc. The Manning grass shrimp (*Thor manningi*) from the Caribbean is a crustacean that is roughly half male, 49% hermaphroditic and only 1% female, while the yellow glasrose (*Aiptasia diaphana*), a sea anemone native to the Atlantic coast of Portugal and into the Mediterranean, is not only trioecious but can also clone itself.

Common ash trees, however, take things a little further still because they can exhibit different combinations of sex along a spectrum, producing seven different sexual expressions of ash tree. There are those that only produce flowers with stamen, those that produce mainly 'male' flowers and a few cosexual, those that produce mainly cosexual and only a few with stamen, trees that are solely cosexual, those that produce mainly cosexual and a few pistil flowers, those that have mainly pistilate and a few cosexual flowers and, finally, trees that only produce 'female' flowers.

This trioecious mating system has often been viewed as an unstable and transitionary state from dioecy to a system of 'male' and cosexual individuals (technically known as 'androdioecious'). But recent studies of the European ash are now questioning this interpretation. What these complex mating systems and sexual organizations show is that generalizations can be tricky to apply to biology, and that language is imperfect when talking about nature's vast sexual diversity and range of combinations.

Common cockchafer
Historical homosexuality

Cockchafers, also known as maybugs or doodlebugs, are a group of three species of beetle in the genus *Melolontha*. Found throughout much of central Europe, they have a long history and deep cultural connection with the region, primarily as a once significant agricultural pest. During the second half of the nineteenth century, they were also the focus of intense scientific debate about the nature of homosexuality. This started in 1834, when the German school teacher August Kelch described his discovery of a larger common cockchafer having sex with a smaller forest cockchafer. What was unusual about this partnering, apart from the fact that it was cross-species, was that both beetles appeared to be male.

It is surprisingly easy to tell the difference between male and female maybugs – males have much larger, fluffier antenna. Despite this, Kelch initially thought that it couldn't be two male beetles and was more likely a pairing between a male and a female with male antenna, perhaps in a similar way to how some female hens and pheasants have been known to develop male plumage. But upon dissection he discovered the truth – these were indeed two male cockchafer beetles. One of the males used its 'penis', known as the 'aedeagus', to penetrate the other through its reproductive opening, pushing the receiving male's aedeagus back into its body cavity. If the antenna hadn't given things away, the act looked like two heterosexual beetles mating.

Kelch and his colleagues tried to rationalize what they were seeing, writing 'thus it was clear that *Melolontha* [*melonlontha*], as the larger and stronger of the two, had forced itself on the smaller and weaker male forest cockchafer, had exhausted it and only because of this dominance had conquered it, so to speak.' This initial notion – that despite the tell-tale antenna, the penetrated individual must have been a female – was echoed again and again over the years as more reports of queer cockchafers

During the 1800s the topic of same-sex pairings among cockchafers enabled entomologists to start questioning why animals have gay sex.

surfaced. And as each of these examples also turned out to be homosexual pairings, researchers frequently fell back on the idea that, somehow, the beetle penetrating must have coerced the other into sex.

Fast forward 45 years and, in 1879, the Russian diplomat and entomologist Carl Robert Osten-Sacken described his own observations of male–male maybug pairings. But this time he questioned the assumption that the sex was somehow 'forced' by a dominant penetrating partner. Osten-Sacken noted that the complexity of the coupling meant that what he would call the 'passive' beetle was clearly willing. He backed this assertion up with the fact that, in a number of cases, the 'passive' beetle was also the larger of the two insects. He argued that if it was all about dominance and strength, how come the smaller beetles were overpowering the bigger ones?

At this point in Osten-Sacken's discussion of queer beetle behaviour, the beetle activity became entwined with conversations about the morality of human homosexuality. Osten-Sacken's use of terms such as 'active' and 'passive' partners, imbued at the time with notions of masculinity and femininity, were presumably

This image of two male cockchafers having sex published in 1896 is thought to be the first ever image of non-human homosexual behaviour.

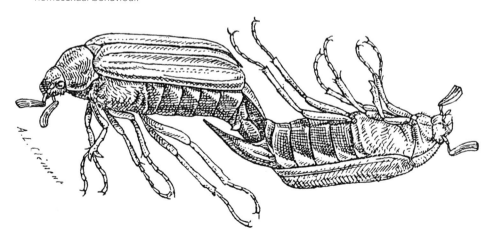

meant to echo the terminology used when discussing sex between men. But despite the suggestion that the beetles were not being forced into queer sex and the parallels he was seemingly drawing with homosexual men, Osten-Sacken carefully stopped short of debating the morality of what he was observing, finishing with the somewhat understated line: 'The interesting philosophical considerations which arise from such abnormal phenomena, I leave to the reader himself to address.'

Things took another step forward in 1896 when the French entomologist Henri Gadeau de Kerville delivered a talk at the Société Entomologique de France conference about the 'sexual perversion' of male beetles. He built on Osten-Sacken's remarks and made the argument that homosexual behaviour observed between male cockchafer beetles came in two forms: necessity and preference. Gadeau de Kerville posited that male beetles kept together would likely have sex out of necessity, but that there was also clear evidence that, even in mixed environments, some males still chose to mate with each other. He even went so far as to say that this behaviour 'also took place among the higher vertebrates', almost certainly a loosely coded reference to humans. This was an incredibly controversial statement to make at the time and was met with fierce resistance. But Gadeau de Kerville didn't back down. He published a long rebuttal in response to this backlash, doubling down on his assertions. In the process he produced an extraordinary drawing of two male cockchafer beetles having sex, thought to be the first ever image of non-human homosexual behaviour. While de Kerville never went so far as to support homosexuality, he was still widely denounced and, on through the 1900s, reports of queer behaviour amongst beetles ebbed away, with those that did appear mainly being published in obscure publications.

This debate illustrates how established scientists continued to categorize queer behaviour in animals as 'unnatural' or 'abnormal'. Despite more and more examples of such behaviour in nature being reported, the legal, social and theological resistance of a society determined to keep the status quo alive held fast, even if it meant twisting and warping observations in order to explain the gay away.

European yew
Sex change

As the last stone was placed on the great pyramid at Giza around 4,500 years ago, a yew tree still alive today in Scotland might already have been hundreds of years old. Over its lifetime, it has seen the bears that once wandered the landscape vanish, Vikings invade and the country of Scotland itself come into existence. The Fortingall yew is thought to have been growing in what is now the small village of Fortingall, Scotland, for up to five millennia, making the tree a contender for the oldest individual living organism in Europe, although this comes with the caveat that it is exceedingly difficult to age ancient yew trees, as the middle of the tree often decays, making it impossible to read its rings.

The tree is a European yew (*Taxus baccata*), a species that belongs to the conifers and is found growing from western Europe to Iran. Typically reaching around 20 metres (66 feet) tall, the trees have small, flat leaves and a thick trunk with thin, scaly brown bark. As the plants age they can become twisted and gnarled, with the trunks often splitting and becoming distorted. In 1796, the Fortingall yew's trunk measured an astonishing 16 metres (52 feet) in circumference before breaking apart into muliple stems. This is one reason why the tree has been so hard to age.

Yews are incredibly toxic. Every part of the plant is poisonous, apart from the flesh that surrounds the seeds, which are eaten and distributed by animals. But it would prove fatal if those same animals were to eat the needles. While most conifers are monoecious, with individual plants having separate 'female' and 'male' structures on the same plant, the yew tree is dioecious. This is a mating system in which the 'female' and 'male' structures are found on separate plants. For the European yew, this means that some trees only produce the pollen-bearing cones whilst others only produce the seed-bearing cones. When fertilized, the seed-bearing cones form little red fruit that bedeck the trees in midwinter and provide tasty little meals for hungry birds.

Every single part of a yew tree is poisonous, apart from the red flesh of the berries, which are consumed by birds and other animals.

For as long as it has been recorded, the Fortingall yew has been considered a male tree. So it was quite a surprise when, in 2015, botanists from the Royal Botanic Gardens in Edinburgh noticed three little sparks of red amongst the deep green on one of the tree's branches. It turned out that the otherwise pollen-bearing tree had developed a single, seed-bearing branch. Why the Fortingall yew suddenly developed a female branch after thousands of years is not known, but it does give us a glimpse of the incredible flexibility of plants, particularly when it comes to sex. One suggestion is that trees can, in effect, create separate entities within the individual plant, as compartmentalizing different parts would give the tree a better chance at fighting disease. Another theory is that this could be a 'sport', where a genetic mutation in one part of the plant gives it different characteristics to the main body. The nectarine, for example, is a sport from the peach that has since been propagated. But it could also be environmental. There are frequent examples of plants changing sex that are also usually classed as dioecious, not just the European yew.

One experiment on the striped maple tree (*Acer pensylvanicum*), usually found growing in the northwestern forests of North America, discovered that subjecting trees to extreme stress triggers a change in their sex. When researchers completely defoliated or severely pruned male maple trees, the trees were four and a half times

The Fortingall yew has always been known to produce only pollen cones, until in 2015 when botanists discovered a single branch producing a handful of red berries.

Yew trees are typically arranged with some having only pollen cones and others only developing seed cones, like these ones.

more likely to grow female flowers than when subjected to less-severe physical trauma. Other experiments have found that restricting the flow of water and nutrients to individual tree branches can also induce a change in the expression of sex in that single branch, when compared to the rest of the tree. It could, therefore, be possible that the single branch of the Fortingall yew that developed seeds may have undergone some form of stress that did not impact the rest of the tree.

But, as is always the case with biology, things are never straightforward. Other trees show the reverse, with female trees exposed to trauma changing their sex expression to become male. What drives the direction these changes take is likely to be individual to the species or groups studied. For example, maples have a highly male-skewed sex ratio, with more than three male trees to each female. With them, therefore, researchers suggest damaged trees with limited time to live would benefit more from developing as females as they would then be more likely to pass their genes on to the next generation.

Regardless of what causes the shifts or the direction they go in, what is amazing here is that despite these plants existing in separate, 'sexed' bodies, they retain the potential to shift their sex. What the future holds for the Fortingall yew is as unknown as the cause of its sexual fluidity so late in life. The three fruits found were picked and, while numerous cuttings of the yew already grow around Scotland, it is hoped that the seeds will eventually be planted and that the tree will produce its first identifiable seed-produced offspring in what could be thousands of years.

European eel
Environment-dependent sex determination

For thousands of years the European eel (*Anguilla anguilla*) has played a significant role in the culture and economy of Europe. Despite this, the natural history of the eel has largely remained a mystery, and even today there are still gaps in our knowledge. One of the biggest of these mysteries has revolved around where they come from and how they reproduce. As the eels seemingly appear out of nowhere in rivers and lakes, the Ancient Greek philosopher Aristotle, who was also a naturalist, first suggested that the fish had no sex and instead spontaneously emerged from the mud. As ludicrous as this sounds today, for hundreds of years no one could come up with a better theory than the pedological origin story.

It was only in the first half of the eighteenth century, when naturalists first discovered adult eels with ovaries, that an eel was identified as female. But the absence of male eels meant that it was still assumed the fish reproduced asexually. This frustrated many academics up to and including Sigmund Freud who, in 1876, spent a month dissecting roughly 400 eels in his search for testicles. Freud's initial foray into eel testicles was fruitless, but undeterred he tried again a few months later and chanced upon a pair of organs hidden away inside the body cavity of an eel that he claimed to be the long sought-after testes. Some 2,000 years after Aristotle, Freud had finally found evidence that eels did not spontaneously grow out of mud and were not asexual. But why did it take such an astonishingly long time to get to this point?

First, we need to understand the extraordinary lifecycle of the eel. While it has never actually been observed, all European eels are assumed to hatch from eggs in the Sargasso Sea off the coast of Bermuda, emerging as tiny, transparent, leaf-shaped larvae. They then drift some 5–10,000 kilometres (3,100–6,200 miles) on currents through the open ocean towards the coast of Europe and North Africa. By the time the

The illegal trade in glass eels is thought to be one of the biggest wildlife crimes in Europe, with billions of dollars' worth of live fish illegally smuggled every year.

Whether or not an individual eel becomes male or female is dependent on a number of different environmental conditions.

eels make it to coastal waters, they have transformed into a more eel-like appearance but retain their transparency. As this point they are now known as glass eels and search for estuaries before migrating up rivers to find ponds, lakes and streams to live in. This triggers the next stage of their development. They gain tolerance for freshwater and pigment to become what are called yellow eels. Depending on factors ranging from food, temperature and sex, the fish usually spend anywhere between 5 and 20 years as a yellow eel. Eventually they transform into their final stage – the silver eel. The exact triggers of this are still poorly understood, but are likely have something to do with the amount of fat they can lay down to power them for the next stage of their life cycle.

The shift to silver eels comes with another dramatic transformation. Their fins become altered, their digestive system shuts down and their eyes physically change size and shape to prepare themselves for their epic final migration. The silver eels swim back down the rivers and out to sea, retracing the steps they made perhaps two decades earlier to the Sargasso Sea where they mate and die. Of course, that's only what we assume happens. Because, despite hundreds of thousands of eels making this journey every single year, no one has yet observed them having sex in the wild.

This complex lifecycle is partly why it has taken so long to figure out how they reproduce. The other reason is their complicated and convoluted sexual development. Up until they become yellow eels, all eels are unsexed individuals and have the potential to become either male or female. When they move into freshwater, some grow ovaries and develop into females, while some develop into an 'intersexual' stage and grow tissue known as the 'Syrski organ'. As this organ contains both testicular and ovarian tissue, it has historically been thought that it can develop into either ovaries or testes, although recent studies have begun to question this. Regardless, the current understanding effectively means that there are two pathways for an eel to become female and one for them to become male. Similar to turtles and crocodiles, it is the environment that determines any individual eel's sex. But whilst things are fairly straightforward for the reptiles, with sex being dictated by temperature (*see* Green sea turtle), things are more complicated for the enigmatic eel. No single environmental condition has yet been identified that determines whether an eel becomes female or male.

The speed at which an individual grows is thought to have an influence, with those that develop rapidly in their early life more likely to grow the Syrski organ and become male. This may allow those that reach sexual maturity quicker to then migrate and mate, whilst it is more advantageous for a female to take longer to grow as producing eggs is more resource intensive. But this is not the only factor. The density of eels in any one environment is also thought to influence sexual development, with those fish living in high density populations being more likely to become male. It has also been suggested that the temperature and salinity of the water in which eels live could influence their sex determination. Whatever the trigger, it is only once the intersex eels have started their migration as silver eels and are swimming in the open ocean that the Syrski organ develops into testes and their male sex becomes clear. It is this circuitous path that male eels take to develop testes that is the likely reason why people were unable to find any male eels for so long.

Understanding eel sex could now play a critical role in their survival. Due to a variety of factors, including the damming of rivers and the over-fishing of glass eels, the European eel is Critically Endangered. The number of glass eels arriving in Europe has declined by up to 95% since the 1980s, and the illegal trade in eels is now the biggest wildlife crime in Europe.

White-throated sparrow
Beyond the binary

The white-throated sparrow (*Zonotrichia albicollis*) is far from ostentatious. This small, brown, black and white creature with a little pop of yellow next to its grey bill is a common migratory songbird, frequenting gardens and woodlands across much of North America. But one of the most intriguing things about this ubiquitous little bird is that it comes in two different colour morphs. Half the population have a black head with a white stripe running down the middle, while the other half have a brown head with a tan-coloured stripe. And the differences aren't just in their physical appearance. The two colour morphs also behave differently. Tan-coloured males are typically monogamous, which means they are more faithful to their partner when nesting; they are also not particularly aggressive when it comes to defending their patch and are poor singers for a songbird. The white-striped males, however, are almost the opposite. They are more likely to be polygamous, mating with multiple partners during the breeding season, and they defend their territories aggressively and give a burst of sweet song that filters through the branches.

These stark differences in behaviour also extend to the females of both colour varieties. What's more, the birds almost never mate within their own colour type. This means that female white-stripe morphs mate with male tan-stripe morphs, and female tan-stripe morphs mate with male white-stripe morphs – opposites really do attract. This extraordinary breeding system means any one individual bird is only able to mate with a quarter of the population. Clearly something curious is happening here. It turns out that the birds have effectively evolved to have four individual sexes.

To understand what is going on with the sparrows, we first need to understand how sex chromosomes work. The DNA in the cells of plants and animals are typically divided up into smaller, distinct sections known as chromosomes. These

The white-throated sparrow is a common visitor to many North American gardens, but it was only relatively recently that scientists understood quite how extraordinary they are.

chromosomes contain most of the genes needed for a creature to function, carrying all the information necessary to make the proteins that build muscle, enzymes, leaves, bone and every other biological matter. How these chromosomes are organized varies across the natural world, but in vertebrates, or those animals with a backbone, most chromosomes form identical pairs. Each unit of the pair contains the same genes in the same order, with a little bit of variation between them. This means that if there is a fault in one of the genes used to build a protein, it is compensated for by its counterpart within its pair. It also means that genes can be swapped between each individual chromosome within that pair.

But some vertebrates have a set of chromosomes that don't match. In humans, these are known as the X and Y chromosomes, while in birds they are the Z and W chromosomes. These parts of the DNA, known as sex chromosomes, can be used to determine sex. Not all creatures (or all vertebrates) have sex chromosomes, but for those that do this mismatch is key, because in most cases it allows for the presence

Due to a chance inversion of a section of DNA containing genes that influence reproductive behaviour, the sparrows have evolved four distinct sexes.

or absence of certain sex-determining genes, and because the chromosomes don't match, they can't swap genes.

When it comes to the white-throated sparrows, birds with two copies of the Z chromosome are usually 'male' while those with one Z and one W are usually 'female'. But peering further into the birds' genetics revealed that, to everyone's surprise, the sparrows were evolving a second set of sex chromosomes. Researchers found that, on one of the chromosomes in the second chromosomal pair, a massive section of DNA is completely reversed. It looked like someone had taken a pair of scissors to a portion of the DNA, cut the section out, flipped it over and put it back in again. In genetics, this is known as a 'chromosomal inversion'. It is thought to happen through pure chance, and usually only affects a small section of DNA containing a few genes. But in the sparrows this section was far longer, with more than 1,100 genes.

Researchers examined which genes this inversion contained and found that it included some related not only to the appearance of the birds but also to their behaviour. It turned out that all tan-striped sparrows had two identical copies of chromosome 2, while all white-striped sparrows had one typical copy and one inverted copy. The fact that the original and the reversed chromosome are so different means that they cannot swap genes between them and so the differences are preserved over time. While this inverted section of chromosome might not contain genes that determine sexual development, it does contain genes that affect the bird's reproductive behaviour. Researchers had found the reason why the two morphs existed and why males and females within each morph very rarely reproduce with each other. This perfectly mirrors how scientists believe the XY and ZW sex chromosomes evolved, and so the researchers concluded that the white-throated sparrows are evolving a second set of sex chromosomes, and effectively split into four sexes: white-striped males, white-striped females, tan-striped males and tan-striped females.

This sounds unbelievable, but the natural world is full of extraordinary ways to organize sex within a species. Two sexes with one set of sex chromosomes might be the most common way but the white-throated sparrow, along with reptiles and fish that have two sexes but no sex chromosomes, lizards with sex chromosome but only one sex, freshwater protozoan with seven sexes and some species of fungi that have tens of thousands of sexes, prove that some species do things differently.

Spotted hyena
Female-led societies

The clitoris is found in all mammal species (and some non-mammals like snakes, ostriches and turtles) and is primarily a sex organ that is made up of the glans (an internal shaft) and two 'crura' or 'legs'. But it is difficult to know the extent to which the clitoris varies from species to species as, to date, so few studies have been conducted on the organ. The spotted hyena (*Crocuta crocuta*) of eastern Africa is a rare exception, as the animals' genitals have been the subject of debate for millennia. The anatomy of female spotted hyenas is unlike any other mammal, making them almost indistinguishable from males. This led to the Greek naturalist and philosopher Aristotle debating whether or not hyenas were hermaphrodites, while the Roman naturalist Pliny the Elder wrote: 'It is the vulgar notion, that the hyaena possesses in itself both sexes, being a male during one year, and a female the next'. Around 1777 Western scientists realized that, while hermaphroditic animals do exist, the hyena is not one of them. Instead, the female has an incredibly elongated clitoris, which is nearly identical in appearance to the male's penis. It grows to be roughly 90% as long as the male's penis and is equally thick. Its structure is such that the urogenital canal runs the length of the clitoris and through the glans at the end, meaning that females not only have to pee through what scientists term a 'pseudo-penis', but also have sex and give birth through it.

The genital mimicry doesn't stop there because the female's pendulous clitoris is also surrounded by erectile tissue, which means that they can – and indeed do – get erections just like their male counterparts. In fact, clitoral erections form an integral part of greeting ceremonies performed when two females meet. During these 'meeting ceremonies', two females approach one another and line up head to tail. One or both then cock their legs and display their swollen erections right next to the other hyena's bone-crushing jaws. When only one individual is erect, she is

The anatomy of the clitoris in spotted hyenas means that females are visually almost indistinguishable from males.

The development of the female's penis-like clitoris is thought
to be related to the matriarchal society in which they live.

usually the most submissive of the pair. Exposing a sensitive reproductive organ to
a peer's sharp teeth is thought to be a greeting display of utmost trust. Not only
that though, females also have what appear to be a scrotum, formed by the fusing of
the labia, which is filled with fatty tissue, meaning that spotted hyena are the only
mammals without a vaginal opening.

The extraordinary structure of the female's genitals begs the question of how the
spotted hyena has sex. The rather complicated affair usually starts with the male
submitting to a female in heat, with the most submissive males being those most
likely to mate. If she is satisfied, the male scoots under the female before she retracts

her clitoris 'much like pushing up a shirtsleeve'. This provides an opening for the male to insert his penis through the clitoris and into the female's reproductive tract. Their sexual gymnastics typically lasts between four and twelve minutes, although it might be repeated a number of times over the course of a few hours. The female's elongated clitoris and fused labia also make birthing difficult. Cubs have to pass through the birth canal, which runs through the clitoris, to be born. This results in the rupturing of the structure and causes extensive bleeding that can take weeks to heal. This traumatic birthing means that around 15% of first-time mothers die in the act, while the fate of the cubs is even more hazardous. As the cubs have to navigate the long and kinked birth canal, as many as 60% suffocate while they're being born. This extraordinarily high mortality rate raises the obvious questions of how and why the females developed a phallus in the first place.

Scientists initially thought that the structure was caused by an influx of testosterone, a hormone belonging to a group known as 'androgens', during gestation, which resulted in the 'masculinization' of the foetus. But experiments in which pregnant hyenas were fed a diet containing androgen-blockers found that females were still born with massive clitorises. So there must be some other process involved – both growth factors and genetics have been suggested but remain untested. Exactly why the structure has evolved is also uncertain. Spotted hyenas live in female-centric clans of up to 130 individuals. Perhaps the complexity of sex, as a result of the phallus, helps explain in part how the females remain in control as, to have sex, a male has to submit to the female. The meeting ceremonies between females could also help maintain a strict hierarchy within each clan.

While no other mammalian species can rival the female hyena and her clitoris, the American black and grizzly bears of North America do, intriguingly, come close. It is thought that 10–20% of female bears in certain populations have a birth canal that runs through the clitoris, in a strikingly similar way to that seen in the spotted hyena. It can be so prevalent that some indigenous peoples have stories about 'male mother' bears giving birth through their 'penis-clitoris'. The extraordinary anatomy of the hyena shows just how flexible animal bodies can be. While it had long been thought that androgens were linked to the masculinization of mammalian (external) genitalia, the hyena clearly contradicts this. There is a lot that we simply don't know about how bodies have developed.

Western gull

Lesbian mothers

In 1972 a pair of researchers set off to study the gulls nesting on Santa Barbara Island, a small island in the Channel Islands archipelago off the southern coast of California. It is difficult to tell male and female western gulls (*Larus occidentalis*) apart, but while studying the birds one of the researchers noticed something curious: some of the couples were sitting on an unusually high number of eggs. Typically, the gulls lay between two and three eggs per nest. But some were incubating up to six or seven eggs, in what the biologists called a 'supernormal clutch'. They concluded that these nests were actually tended by two female birds that had formed a lesbian pairing, courted, laid eggs and raised young together.

Studying the colony of western gulls for three years, they found that out of the 1,200 gull pairs studied, up to 14% were queer. It is important to mention here that some of the eggs were fertile, which means that around 15% of the female gulls were consorting with males at some point before returning to their female partner. So while we can't say that the gulls are 'lesbian', we can say that they form lesbian pairings. By 1977 the scientists were eventually ready to publish their findings, but they couldn't have predicted the political fallout that resulted.

The research dropped right into the middle of the gay liberation movement in the USA. A decade after the Stonewall riots, queer communities across the country were organizing themselves and confronting the oppression they were experiencing. In the same year that the gull paper was published, gay politician Harvey Milk was elected as city supervisor of San Francisco. At this time there was very little public awareness of queer behaviour in animals, and one of the prevailing arguments against homosexuality was that it was 'unnatural' and so 'against God's will'. The lesbian gull research, at that moment in time, was dynamite.

The western gull is found all along the west coast of North America, from Canada in the north down to Mexico.

The research on 'lesbian' gulls came at a time when queer communities across the USA were organizing themselves and challenging the status quo.

The research sparked countless newspaper articles and satirical cartoons. One headline ran 'Your tax $$ wasted to study gay gulls', while a group out of New York published a statement claiming that '100% of the sea gulls in the five boroughs of New York City were heterosexual', presumably trying to imply that the queer gulls were a Californian quirk. It didn't stop there. When the House of Representatives debated the National Science Foundation's budget in 1978 (the NSF had partly funded the research), the gull research was referenced and the budget held up for 10 days.

Despite the protestations of the religious and conservative right, the scientists were not put off. The team received countless letters of support from gay men and lesbian women and looked next at colonies of ring-billed (*Larus delawarensis*) and California gulls (*Larus californicus*) to assess the frequency of supernormal clutches and, by extension, queer behaviour. Publishing a paper in 1979 they found a much lower rate of between one and two percent of lesbian pairs in ring-billed and a single polygamous group of one male and three females in California gulls. The study became something of a cultural touchstone, inspiring a song titled *Lesbian Seagull,* written by Tom Wilson Weinberg in 1979 and later covered by Engelbert Humperdinck for the 1996 animated comedy film *Beavis and Butt-Head Do America*, and in the same year a Los Angeles theatre put on a play titled *Supernormal Clutches.* The discovery of lesbian pairings amongst the western gulls might have been the first time a large proportion of the American society came across queer behaviour in animals, thanks to the level of media and political attention it received.

Generations of farmers may have already been well versed in the queer behaviour of livestock (*see* Domestic sheep), but not the general public as scientists appear to have previously been unwilling or unable to publish observations. In the wake of the western gull paper, however, there followed a slow but steady flow of reports in scientific literature of homosexuality in the natural world, including in many other species of seabirds. Over the next few decades numerous papers discussed the queer behaviour of countless gull species, from black-legged kittiwakes (*Rissa tridactyla*) to laughing gulls (*Leucophaeus atricilla*), as well as studies of several tern species, such as the roseate (*Sterna dougallii*) and Caspian terns (*Hydroprogne caspia*) that form same-sex couples. Lesbian pairings have also been found in Antarctic petrel (*Thalassoica antarctica*) and Cory's shearwater (*Calonectris borealis*), which represent the first known species of burrow-nesting seabirds to show queer behaviour.

Perhaps the most well-known queer pairings in the bird world are those of the albatrosses. A number of albatross species form lesbian couples, but one of the most studied is the Laysan albatross (*Phoebastria immutabilis*). Around 31% of Laysan albatross pairs on Oahu, Hawaii, are between two females, with one female sometimes mating with a male before going back to her same-sex partner to raise the chick. Why queer behaviour is seemingly so prevalent in seabirds is not known. It could simply be a sampling bias, as seabirds are a fairly well-studied group of birds, or it could relate to their colony nesting antics, which allow scientists to easily observe a large number of birds and keep track of relationships. Regardless, the story of the western gulls shows how science is inherently political insofar as it relates the natural world with human behaviour.

Common bottlenose dolphin
Explaining the gay away

A few different species of dolphins are known to engage in queer sex, but it is best known in the bottlenose dolphin (genus *Tursiops*). Technically made up of at least two species, the common and Indo-Pacific, bottlenose dolphins are found swimming in warm and temperate waters all around the globe. As the most common species of cetacean in captivity, they are one of the most studied, and this extends to their homosexual behaviour. Both females and males house their genitals in what is known as a 'genital slit'. This is exactly how it sounds – it is a pocket within which they keep anything that might cause drag when swimming at high speed, so variously includes their vulva, penis, nipples and testicles.

Male couples are an ubiquitous feature of bottlenose dolphin communities. When they are young, males typically pair up and remain together for their rest of their lives. They travel together, guard each other, and even fight off sharks to protect one another. They also engage in copious amounts of gay sex. During the first half of their lives, most male dolphins hardly mate with the opposite sex, which means that most sexual activity is within their gay couple. This includes everything from simple nuzzling and courtship behaviour to inserting their flippers into each other's genital slits to stimulate their penis and full-on penetration, both of their genital slits and occasionally their anus.

But this behaviour isn't limited to the males. Female bottlenose dolphins also engage in plenty of lesbian behaviour, inserting their flippers into each other's genital slit to stimulate their large clitorises. The closely related spinner dolphins (*Stenella longirostris*) go one step further – females have been observed riding each other's dorsal fins, with one inserting its dorsal fin into the other's genital slit before both go for a little stimulating swim.

This isn't the only form of propulsive penetration that dolphins undertake. They also use their sensitive snouts to give each other a kind of 'oral' sex, pushing their

Both male and female common bottlenose dolphins engage in queer
sex and behaviour, ranging from simple nuzzling to full on penetration.

beaks into the others' genital slit. Whilst still in this position, the dolphin then starts
swimming, taking their partner with them in a behaviour known earnestly as 'beak-
genital propulsion'. In some cases, the penetrating dolphin even rotates as its swims,
presumably to give their partner the best possible experience.

Despite this seemingly unambiguously queer behaviour, scientists have still
historically tried to explain it away as something else. Determined not to accept
that bottlenose and spinner dolphins display gay sexual activity, researchers have
claimed that such behaviour was 'proof' that intercourse was divorced from sexual
intentions and was instead 'greetings' and 'social communication' (even if that sex

The breeding behaviour of redshanks has been reclassified over
and over, seemingly in an attempt to avoid referring to homosexual
behaviour as sexual.

was between a male and female). By some accounts, in up to a quarter of the cases
where homosexual behaviour in animals has been observed, it has been classified as
something other than sexual.

Redshanks (*Tringa totanus*), for example, are birds that chase each other during
breeding season. Initially, researchers thought the birds in pursuit were of the same
sex, and classed the chasing behaviour as a non-sexual, territorial act. When they

discovered that the birds were actually the opposite sex, they reclassified it as part of sexual courtship display. When some researchers realized that it was sometimes occurring within same-sex pairs, they flip flopped once more and claimed that only the homosexual behaviour was non-sexual (even after the two males copulated). Even up to the 1980s, one author of a book about orangutan behaviour wrote: '… two males [orangutans] regularly mouthed the penis of the other on a reciprocal basis. This behaviour, however, may be nutritively rather than sexually motivated.'

Queer behaviours have been difficult to recognize because of how behaviours have been explained and because of the language used. Over the past few hundred years many scientific papers published in prestigious journals by scholarly academics have used charged language to describe homosexual behaviour. Scientific literature is full of morally loaded terms such as 'immoral', 'bizarre', 'abhorrent' and 'paradoxical' when it comes to describing homosexual behaviour, while gay sex is also frequently referred to as 'pseudo-copulation' or 'sham mating'. An extraordinary example of this value-laded judgement is in the title of a paper from 1987, describing queer behaviour in Mazarine blue butterflies of Morocco: '*A Note on the Apparent Lowering of Moral Standards in the Lepidoptera*'.

This is hardly surprising when homosexuality between men and between women has, in some countries, been illegal and denigrated for hundreds of years. The science often reflects the attitudes of society. What these examples really show is that science is influenced and informed by people, and people are packed full of explicit and implicit biases, which have resulted in centuries of obfuscation, disinformation and coverups. While the science around queer behaviours in the natural world has become more objective in the last few decades, it is still often far too easy to slip back into old prejudices.

Common pill woodlouse
Bacteria-dependent sex determination

Found in gardens and parks across Europe, the common pill woodlouse (*Armadillidium vulgare*) is a familiar creature to any child who has rummaged around the leaflitter or under moist logs. A crustacean that made the bold life decision to move from water to land, the woodlouse is hiding an incredible secret: its sex is determined by bacteria.

In many animals, sex is often determined by sex chromosomes. In most mammals these are the XX and XY chromosomes, with males typically having XY and females typically XX. Other species have ZZ and ZW chromosomes, in which the females typically have ZW and the males typically ZZ. Because this system of determination is so common across the natural world, it has often been assumed that the sex chromosome's role in such a fundamental developmental process was unlikely to vary. But this is far from the case. Countless times in every branch of the evolutionary tree, scientists have found that animals have been tinkering with their sex chromosomes, from pulling bits out, putting other parts in, creating new ones to getting rid of them entirely.

The common pill woodlouse falls into the ZZ and ZW sex chromosome camp, with ZW females and ZZ males; yet many females will only produce other females. This is because, like most arthropods (animals with no backbone), the woodlice are vulnerable to infection by a bacteria known as *Wolbachia*. First identified in 1924, *Wolbachia* is a genus split into about eight highly diverse supergroups, and while all of these supergroups are currently thought to form a single species called *Wolbachia pipientis*, scienists have proposed other species within the genus.

Wolbachia is a parasitic bacterium that most commonly infects the reproductive organs of arthropods and is only passed from a female to her offspring (whereas normally an infection would be passed between all individuals in a population). The female's larger eggs are thought to have more cytoplasm to support the hitchhiking parasite. In woodlice, this had led to an extraordinary situation. Because the bacterium is passed on by the female, *Wolbachia* fares better if there are more females

In some populations of common woodlice, infections of *Wolbachia* have meant that there are no longer any genetically 'female' individuals.

in a population. As a result, the bacterium has evolved to influence male woodlice embryos to develop as females. This means that genetically ZZ males are physically females that go on to lay infected eggs, ensuring the continued spread of *Wolbachia* down the generations.

In some populations of woodlice the *Wolbachia* have been creating ZZ females for so long that it has resulted in the loss of the W chromosome entirely, meaning that there are no genetic females. This means the bacteria completely controls sex,

Studying the woodlouse revealed an entirely new way for sex chromosomes to evolve – through the horizontal gene transfer from bacteria.

with uninfected ZZ individuals developing as males and infected ZZ woodlice developing as females. There has been a shift from chromosomal sex determination to cytoplasmic sex determination.

Yet the story doesn't stop there. In 1984 scientists studying the woodlice found something else. In some populations they discovered that there were uninfected ZZ females. This made no sense, as they needed the influence of the *Wolbachia* to prevent them from reverting to males. It turned out that the section of DNA in the bacteria responsible for creating the ZZ females had, in effect, broken off and been reinserted into the woodlouse's own genome. While *Wolbachia* had originally been responsible for causing the woodlice to lose their female sex chromosome, it now looked as if they were also responsible for creating a new one. This incredible discovery showed just how flexible genetic systems are and provided an entirely new route for how sex chromosomes can evolve.

This extraordinary relationship between the *Wolbachia* and the humble woodlouse is far from unique. *Wolbachia* is thought to be the most common reproductive parasite in the world, infecting up to 66% of all arthropod species. The likelihood is that, right now, there is an insect crawling around your house carrying the bacteria. This has resulted in lots of other species of arthropods also evolving a close relationship with the bacteria. In several species of parasitoid wasps, for example, the *Wolbachia* has effectively eliminated the males in a population by causing unfertilized eggs to develop as female wasps. This occurs as *Wolbachia* exploits the typical way in which wasps determine sex.

Normally, male wasps develop from unfertilized eggs, which only have a single set of chromosomes, while females develop from fertilized eggs that have two sets of chromosomes. *Wolbachia* bacteria cause chromosomes in unfertilized eggs to duplicate and so turn the resulting wasp from male to female. Studies have shown that if infected eggs are cured of the bacteria they continue to develop as males, but if these males then try to mate with females, they are unsuccessful. It has been suggested that populations have been parthenogenic for so long that the part of their genome responsible for sexual reproduction may have been effectively corrupted. As a result some of the species of *Trichogramma* wasps may also no longer be able to reproduce uninfected.

The impact of *Wolbachia* on the invertebrate world cannot be overstated. It has even been suggested that the bacterium now plays a role in the creation of new insect species, as it has reproductively isolated certain populations. Scientists are attempting to harness the bacteria to control malaria-carrying mosquitos. The extraordinary thing about *Wolbachia* is how one living organism is able to control and influence the sex determination system of a completely unrelated species. Because it is so successful at this, it has evolved to alter the sex of a hugely diverse range of species across massive swathes of the animal kingdom.

Bluegill sunfish
Do animals have gender?

Throughout the scientific literature scientists have used the terms 'sex' and 'gender' interchangeably. Whilst the terms might once have been synonymous, language has changed and they are now distinct, with sex often referring to chromosomal, hormonal or gonadal sex and gender an innate sense of self in relation to socially constructed norms. In the same way that the terms 'gay' and 'lesbian' are typically restricted to humans, as we can never know whether animals are, we can clearly never know if animals have a sense of gender or what that would mean even if they did. But a number of species of plants and animals – such as the bluegill sunfish (*Lepomis macrochirus*) – can help question our assumptions about how the natural world balances this contrast between an individual's internal sex and their external presentation. They also expose a host of human biases when an animal does not match the preconceived notions of how males and females 'should' behave.

The bluegill sunfish is a common species of freshwater fish found in the streams, rivers, lakes and wetlands of North America. Growing to around 30 centimetres (12 inches) long and weighing up to 2 kilogrammes (4 pounds), the fish typically breed from late May to mid-August, during which time large males excavate and defend a nest in the gravel or sand on the lake bottom. Females then approach the males and perform a courtship dance in which the fish swim around each other while the male aggressively harries the female. If the female is satisfied she lays her eggs in the nest as the male releases sperm to fertilize them.

What is interesting with the bluegill is that the males come in three different forms, separated by size: the largest are the nest-building males, which have an orange tint to their chests; the intermediate-sized males feature cryptic dark striping on their sides; and the much smaller males have no discernible colouring at all. While the larger males are busy courting the females and encouraging them to spawn, the small males dart into the fray and release their sperm indiscriminately in an attempt to fertilize as many eggs as possible. Thus the large males must defend themselves against not only other large males, who might want to take over their nest, but also the smaller males. When

The largest male bluegill sunfish build nests in the sand at the bottom of lakes and rivers in an attempt to win the attention of passing females.

the small males eventually develop into the intermediate males, things get even more interesting. Intermediate males look like females and enter the nests of the large males without eliciting any aggression. Instead, the larger males allow them to take part in the courtship routine and even to fertilize some of the eggs. The intermediate male displays a behaviour described in scientific literature as 'sexual mimicry'.

Bluegill are not alone in having multiple types within one sex. The giant cuttlefish (*Sepia apama*) and Broadley's flat lizard (*Platysaurus broadleyi*) have female-presenting males and male-presenting males, while the European ruff (*Calidris pugnax*) and the marine isopod (*Paracerceis sculpta*) both also have three different types of male and one female, similar to the bluegills.

The reverse is also seen. In the blue-tailed damselfly (*Ischnura elegans*), the males are typically an electric blue and the females a dramatic red, but in some populations up to a third of all females are blue-coloured. It is thought that, in populations with large numbers of males, the females present as males in order to avoid being harassed by the males who might otherwise want to mate.

Whether or not these different types of males and females could be called different 'genders' is open to debate. Evolutionary biologist Joan Roughgarden plays with the idea that 'gender is the appearance, behaviour, and life history of a sexed body' and so refers to these animals as having multiple genders. This is not, however, a universally accepted definition. Regardless, scientific literature often uses loaded language when describing sexual mimicry. Frequently, the smaller and female-presenting males are described as 'sneaky' and 'deceitful', and scientists claim that they are

The intermediate-sized male sunfish are thought to look like female fish so that they can enter the nests of the large males without provoking any aggression from them.

'stealing fertilizations' from the male-presenting males or engaging in 'cuckoldry' in comparison to the larger male's 'bourgeois' behaviours. These descriptions often seem to imply that the larger males are owed the right to mate with the females. For example, the smaller and intermediate male bluegill fish are described as displaying 'alternative' mating strategies, when in fact they make up 80% of the breeding males. In this case, the strategies of the small and medium males are clearly the norm, and it is the larger males that are engaging in the 'alternative' way of life.

Does any single male have the 'right' to mate with a female? If the answer is no, then why do some researchers and documentaries describe female-presenting males as 'sneaky' and 'stealing fertilizations'? It seems possible that the application of moral language onto these behaviours is rooted in the notion that the male-presenting males are somehow the ones doing things 'correctly'.

The smallest males only dart in to spray their sperm when the larger and intermediate males have successfully courted a female.

Common pheasant
Out-sized influence

Historical literature is, by and large, free from mentions of queer natural history but, curiously enough, there were ardent discussions about 'sex-transformative' birds in the late eighteenth and nineteenth centuries. These discussions paved the way for scientists to explore traditionally taboo subjects such as homosexuality and the changeability of sex within humans.

The Scottish naturalist John Hunter set the ball rolling in 1780 with the publication of a paper called *An Account of an Extraordinary Pheasant*. In it he documented the case of a female common pheasant (*Phasianus colchicus*) that shed its feathers and replaced them with male plumage. Reports of birds reversing their plumage were uncommon, but those people who worked closely with domestic and game birds certainly knew of them. No-one had an explanation for what was termed incorrectly at the time as a 'sex-reversed' pheasant, and Hunter immediately began to compare it to humans. He wrote: 'We find something similar taking place even in the human species: for that increase of hair observable on the faces of many women in advanced life, is an approach towards the beard, which is one of the most distinguishing secondary properties of man.'

At this point Hunter touches upon one of his more influential and long-lasting notions, as his mention of 'secondary properties' is a direct forebearer to what are today known as 'secondary characteristics'. These features, such as the antlers of a deer or the extravagant tail of a peacock, are usually typical to one sex but not directly related to reproduction. These secondary properties inspired Charles Darwin's second theory of sexual selection, and in fact the ability for the female hens to seemingly change sex also fed into the foundations of his theory of evolution. This theory relied on the understanding that life was flexible, able to change and adapt to different environments, and the mutability of sex appeared to support this.

We can never be certain why the female birds that Hunter studied started to present as males, but we do know of a potential cause. Within the group of birds known as Galliformes, which includes chickens, turkeys and pheasants, the male

Some of the earliest mentions of queer observations in natural history involve female pheasants developing the plumage more typical of males.

plumage is the default patterning and, in female birds, requires oestrogens to suppress this and develop female-typical plumage. Experiments have shown that if a female's ovaries are damaged, oestrogen levels drop and result in the dramatic development of male feathers.

But these are not the earliest accounts of female birds presenting as males or vice versa. To find that we need to go back a further 2,000 years. Aristotle is largely known as an Ancient Greek philosopher who delved into metaphysics and epistemology and tutored Alexander the Great. But he was also a naturalist and, by many accounts, one

If the ovaries of a female pheasant are damaged, it can cause their levels of oestrogen to drop and result in the dramatic development of male feathers.

of the first recognizable biologists. He was also one of the first people in Western science to document non-heteronormative animals. Aristotle wrote about what he thought were 'hermaphrodite' hyenas *(see* Spotted hyena) and considered the eel to be an asexual fish that spontaneously generated from the mud (*see* European eel). While he was wrong about both of those things, he also talked about birds that had 'reversed' their sex. In his book *Historia Animalium*, Aristotle described how some domestic hens 'crow in imitation of the males and attempt to tread, and their crest and tail are raised so that one would not easily recognize that they are females; in some there has even been an outgrowth of a sort of small spurs.' While he also noted that in some cocks 'there are also some birds that are effeminate from birth to the extent that they even submit to males attempting to tread them.'

COMMON PHEASANT

This ability of some animals to present as their opposite sex is not limited to the avian world; numerous examples come from across the animal kingdom. Some species have built this into their breeding system (*see* Bluegill sunfish). In plenty of other cases they appear randomly in populations similar to the birds described by Hunter and Aristotle. For example, there have long been reports of female deer growing antlers. As far back as 1724 the Dutch naturalist François Valentijn included in his book on the history of the Dutch colonization of Southeast Asia, *Oud en Nieuw Oost-Indiën*, a drawing of 'A horn from a Ternaatsche hind', presumably in reference to a doe from around the city of Ternate in Indonesia. In the modern day there are numerous examples from North America of hunters killing antler-bearing white-tailed (*Odocoileus virginianus*) and mule does (*Odocoileus hemionus*), with the Maryland Department of Natural Resources estimating that this occurs in roughly one out of every 20,000 females.

Another well-known example is of a group of five lionesses in southern Africa. Male African lions (*Panthera leo*) are typically distinguished by the large shaggy mane that runs from the tops of their heads, over their shoulders and onto their chests. But there have been reports of female lions growing manes – in 2016, a paper was published documenting five male-presenting lionesses from the same protected area in Moremi Game Reserve, Botswana. Intriguingly, like the reports of the hens, these lions don't only look but also act like males. For example, one of the male-presenting female lions known as SaF05 shows what would be considered typical male behaviours, such as scent-marking and increasing the amount she roars, as well as mounting other females in the pride.

While they may not be 'sex-reversed' or 'hermaphrodite', as originally thought, examples of these male-presenting female animals have had a little-known but significant impact on natural history, inspiring some of the most influential scientists of all time.

Splitgill mushroom
Thousands of sexes

Most multicellular life on Earth reproduces in a remarkably similar way. From cacti to capybara most organisms use a system of sexual reproduction in which individuals produce a large sex cell, a small sex cell or both, a process known as 'anisogamy' (with all the caveats that highlight the numerous asexual plants and animals that we know exist). But this large/small view of sex overlooks a hugely significant part of multicellular life: the fungi. Fungi are an extraordinarily diverse part of life on Earth. They can be found growing in every environment, from the icy waters of Antarctica to rocks on the highest mountain peaks, from the surfaces of leaves in tropical forests to baking deserts. A teaspoon of Amazonian soil is thought to contain as many as 400 species of fungi. But while just 150,000 species of fungi have been described to date, it is estimated that this only represents up to 10% of all the species of fungi that exist, with one study even suggesting that there could be as many as 3.8 million species. Considering that various estimates put the number of all species on Earth at around 8-10 million, it is possible that fungi could represent almost half of all multicellular life on this planet.

And when it comes to sex, fungi do things in hugely diverse ways. While not all fungi species reproduce sexually, many of those that do are in fact 'isogamous', which means that all the sex cells are identical in size. This means it is tricky to apply words such as 'sex' or 'male' and 'female', and scientists usually talk about fungal 'mating types' instead. Species such as the splitgill mushroom (*Schizophyllum commune*) have evolved more than 23,000 mating types. This attractive species of mushroom can be found all around the world and is distinguished by tightly packed, pinkish-cream fans covered in light fur. But to understand how their 'sex' works, we need to take a dive into the complex world of fungal sex.

The splitgill mushroom has mating types determined by two different loci that come in a variety of different forms, resulting in 23,328 'sexes'.

Fungi are incredibly diverse but vastly understudied organisms, with some studies suggesting they could account for almost half of all multicellular species on Earth.

Broadly speaking, fungi can reproduce in one of three different ways: asexually, sexually or parasexually. With asexual reproduction, the organism can fragment, bud or produce spores. Fragmentation means that the root-like hyphae of the fungi break apart and then regrow as an individual organism, while budding occurs when a cell bulges from one side until it literally buds off and produces a new individual. But the production of spores tends to be the most common way for fungi to reproduce asexually. All asexual methods mean that offspring are pretty much identical to the parent. Many fungi happily reproduce asexually but then switch to sexual reproduction if the environment changes. Sexual reproduction and mating compatibility in fungi is largely controlled by the genes found at a point (known as a locus) on the genome known as the 'mating type', abbreviated to 'MT'. These genes can produce different variants of a protein, meaning individuals with differing mating types are needed for successful sexual reproduction. Roughly speaking, this can be thought of in similar terms to the genes on the X and Y chromosomes in mammals, and so mating types are, to a certain extent, analogous to sex.

To make things more complicated, sexual reproduction in fungi is then further broken down into two main types. The first of these is known as 'homothallic', homo meaning 'same' and thallic referring to the tissue of the fungi. This form of sexual reproduction is most similar to co-sexual plants or hermaphroditic animals, where individual organisms have both the male and female reproductive structures within a single body. Within fungi, this means that individual organisms are capable of expressing both the MT variations within a single cell, allowing the fungi to reproduce with itself. This is contrasted with 'heterothallic', meaning 'different' tissue fungi. These species require two individuals from different mating types to share some of their genome that combines to form the offspring. At the most basic, these individuals might have an MT controlled by a single locus with two variants which results in two different mating types. But there is more.

Some fungi have an MT determined by a single locus that has multiple variants resulting in dozens of mating types. Others, like the splitgill mushroom, have mating types determined by two loci. These two loci can then come in a variety of forms. In the splitgill mushrooms the first can be one of 288 different types and the second one of 81 different variants, meaning that any individual mushroom can be one of 23,328 possible mating types, or 'sexes'. Splitgill fungi contain a huge amount of genetic variation, guaranteeing that any individual can mate with any other fungi it comes across.

Finally, some species of fungi use parasexual reproduction. This occurs when the root-like hyphae of individual fungi touch, and two cells from each organism fuse. The DNA from the unrelated fungi come together and exchange genes before the cell splits apart to form new individuals. A lot of the steps in this process are similar to sexual reproduction, including the exchanging of genetic material, but this completely separate system has evolved independently and is found only in fungi and single-celled organisms.

This is a mere glance at the variations, nuances and subtleties of fungal sex. For example, some species not only have a variety of mating types but also different-sized sex cells, adding another level of complexity to the system. The huge range of fungal species showing equally kaleidoscopic variations of reproduction is almost endless and still not fully understood, but it helps demonstrate yet again the sheer diversity of life on Earth.

Chinese shell ginger
Temporal sex

Across the natural world, where species have different sexes, those sexes are often found in different bodies. But that is not always the case. A number of species of animals are known to be simultaneous hermaphrodites, while flowering plants commonly produce cosexual flowers. But some species of tropical ginger have done something different again. They effectively separate their sex not spatially, but temporally.

The Chinese shell ginger (*Alpinia kwangsiensis*) typically grows in the tropical seasonal rainforests of Yunnan province, southern China, where it inhabits sunny gaps on the forest floor and on the sides of roads. The plant is a small herb, which means that it has a soft, fleshy stem, as opposed to a woody trunk, that grows up to 3 metres (10 feet) tall. Beautiful, long, blade-like leaves stick out along the length of the stem until its end, at which point a cluster of spike-like flowers emerges. The flowers only last for a day and resemble a slightly squished cone, pinkish-red on the inside and creamy-white on the outside. In the centre of each flower is a yellow-coloured structure.

The flowers are technically cosexual. The yellowy structure contains the pollen-producing stamen and the egg-producing pistils. But this creates a problem. If the plant were to produce flowers that have both mature pistils and stamen at the same time, it runs the risk of fertilizing itself. Plants of all different shapes and sizes have arrived at their own way of avoiding this situation, such as putting the 'female' flowers on one plant and the 'male' ones on another (*see* European yew); but others tightly control the time at which they produce pollen and eggs instead.

When the cluster of flowers open in the morning, the stamen are positioned in a such a way that they hang down, beneath the pistil. This allows the anther to produce

The species of Chinese shell ginger *Alpinia kwangsiensis* arranges its sex temporally rather than spatially as it switches from one to the other during the day.

All jack-in-the-pulpit plants start out as 'male' but transition
to 'female' when they get older.

pollen to its heart's content, as large, solitary bees buzz in and out of the flowers and spread its pollen far and wide. The flowers are effectively acting as a 'male' at this point but, as soon as it hits midday, everything changes. The physical structure of the flower alters, as the anthers stop making the pollen and the pistil descends. The style hangs lower than the stamen and becomes receptive to pollen, which can then go on to fertilize the eggs. At this point, the flower has effectively become 'female'. But if all the flowers on each plant follow this set routine, there is a problem. Obviously if all flowers start out as male in the morning and transition to female in the afternoon, no plant would ever get fertilized. To counter this, roughly half the ginger population does things the other way round, starting off as 'female' in the morning before the style lifts up and the anthers descend to become 'male' in the afternoon. That way the busy bees can transport the pollen to fertilize flowers between plants but not within.

And the Chinese shell ginger is not alone in doing this. Researchers have discovered that eight other species in the same genus produce flowers that change sex during the day, and they suspect there may be more waiting to be discovered.

There are numerous other examples of similar tactics elsewhere in the plant world. The jack-in-the-pulpit (*Arisaema triphyllum*) is found in the moist, temperate forests of southern Canada and eastern USA. It is a member of the arum family, a group of flowering plants that contains some 4,000 species and is perhaps best known for the peace lily (*Spathiphyllum cochlearispathum*). The group's most distinguishing feature is the striking flower spikes, called a 'spadix', which are surrounded by a bowl-like modified leaf known as a 'spathe'. While we tend to think of these spikes as an individual flower, they are actually a collection of many smaller flowers. These flowers are found at the base of the spadix within the confines of the protective spathe, with the long, slender visible portion modified to produce heat and an odour to attract insect pollinators.

When the jack-in-the-pulpit puts out a bloom, the flowers it produces are either all male, all female or a mixture of the two. Which of these it develops depends on how much starch the plant has stored, with the size of the plant linked to its age. The plants all start out as male, creating flower spikes that only develop pollen-producing anthers. Every year the plant's leaves die back, leaving behind a starch-storing tuber that fuels the next year's growth, with the plant transitioning as it ages to produce a mix of male and female flowers until it reaches its final stage. The oldest and largest plants produce a spadix with only female flowers, which once fertilized turn into a collection of bright red berries. The plant with the world's largest flower spike, the titan arum (*Amorphophallus titanum*), produces both male and female flowers, but goes in the opposite direction as, for the first few days, it only opens the female flowers before switching to opening the male ones.

While not often seen in plants, this temporal shift in sex demonstrates yet again the vast range of different ways in which the sexed body is organized in the natural world and the ingenious ways in which genetically diverse offspring are produced.

Cane toad

Intersex animals

Every so often a news article surfaces about chemicals in the waterways causing amphibians to change sex. Often reported as a commentary on the growing human impact on the natural world, these stories actually obscure an inherent ability that many frogs and toads (collectively known as the 'anurans') possess to blur the boundaries between the sexes.

Most male 'true' toads, or those that belong to the family Bufonidae, have what is termed 'ovotestis' and so can produce both sperm and oocytes, which are effectively the precursors to egg cells. These toads include the cane toad (*Rhinella marina*), a large amphibian native to South and Central America, where it plays an important role in the ecosystems in and around the numerous waterways that crisscross the tropical environment. The toads are, perhaps, more notorious for their introduction into Australia during the 1930s in an ill-fated attempted to control beetle pests threatening sugar cane crops. The experiment did not work, and the toads wreaked havoc on the native Australian wildlife. This was clearly no fault of their own, but it does mean that, because they have been so well studied, we now know quite a lot about these amphibians.

Along with most other toads, cane toads have what is known as a 'Bidder's organ'. This piece of tissue sits just in front of the kidneys and is derived from the same gonadal tissue which, in females, goes on to form the ovaries. In males, half of the tissue develops into testes while the other half retains connective tissue, blood vessels, oocytes and egg follicles in various stages of development. The role of this organ is still not fully understood, but it might be linked to hormonal regulation. Scientists have found that if male cane toads are castrated – either manually or chemically – the Bidder's organ kicks into action and they start to develop egg cells. While

Most true toads, such as this cane toad, have a particular organ called the 'Bidder's organ' which in males contains miniature egg follicles that kick into action if the toad is castrated.

It was thought that human hormones might be turning male green frogs female, but one study found that this occurs even in remote populations.

the male toads cannot lay these eggs, experiments have suggested that the origin of the Bidder's organ and the ovary are identical, meaning that all male toads would be what biologists consider 'intersex', although the presence of the organ in female toads has recently questioned this origin.

While not all anurans have a Bidder's organ, that doesn't stop many from showing a fluidity of sex, though until recently direct evidence of frogs naturally changing sex in the wild has been rare. The green frog (*Rana clamitans*) is a large, chunky species, typically found wallowing in the lakes, swamps and freshwater ponds of eastern North America. Studies have found that male green frogs with testes that produce egg-like cells are quite common in ponds in and around suburban developments, while there is also a shift in the ratio of males to females. This has led many to suggest that waterways heavily contaminated with human chemicals, such as oestrogen and pesticides, are disrupting the normal hormonal pathways in male frogs and causing a shift in the sexual development of tadpoles. But that is not the full picture. Scientists have also looked at populations of wild green frogs living in ponds far from any perceived human influence and, to their surprise, found that sex reversal and intersex males in green frog populations 'may be common and frequent'. The results showed that between 2–16% of all individuals were either intersex or showed evidence of having switched their sex entirely.

Despite previous assumptions that these reversals were largely driven by human activities, it seems likely that, instead, they are occuring naturally, undetected within amphibian populations. The determination of sex within frogs may therefore be a combination of both genetic and environmental factors we are yet to fully understand.

While the Bidder's organ in toads means that it is the males that possess the ovotestis, in some species of moles it is the females. Moles of the genus *Talpa*, which includes the European mole (*Talpa europaea*), along with a few other mole species such as the star-nosed mole (*Condylura cristata*), are the only mammal species in which all the females possess ovotestis. Within these burrowing mammals, the ovarian tissue is fully functioning and produces egg cells, while the testicular tissue is unable to produce sperm cells but does create high levels of testosterone.

The presence of the ovotestis even causes the external genitals of young female moles to look almost indistinguishable from those of males, to such an extent that professional mole catchers often reported that all moles were male until one year old, at which point some would become female when the vaginal opening appeared. The reason why the female moles possess ovotestes that pump out testosterone is thought to be behavioural. Living underground and digging for worms for most of the year takes a lot of energy. Researchers found that during the non-breeding season, the testicular side of the ovotestes expands so much that it becomes larger than the ovarian side, providing a massive boost of testosterone that not only helps females burrow but also defend their little patch from potential intruders.

The cause of ovotestes development in females is an alteration in the regulation of genes that produce testes. While we tend to think of the development of sex as an on/off switch dependent on the presence or absence of certain genes and chromosomes, it is becoming increasingly clear that a complex network of genes can provide alternative pathways for the development of sex.

Moss mites
Ancient asexuals

Evolutionary biologists have long struggled to explain why the vast number of multicellular species on this planet reproduce through sexual reproduction. While sexual reproduction offers the chance for species to mix their genes up and produce variation that can be helpful if the environment changes, scientists have also noted that it comes at significant cost.

For a start, a species has to find a mate, which can expose individuals to the risks of predation and sexually transmitted diseases. Genetically it can also cause problems by risking the break-up of groups of genes that work well together. Finally, and perhaps most significantly, it means that only half a population can actually produce offspring. Surely it would be much more efficient if every single animal could just have babies on its own? Some scientists have suggested that sex is more beneficial in the long-term, because asexual plants and ani mals are 'doomed to early extinction'. And, in fact, many species of asexual creatures are thought to be relatively newly evolved. But, as is often the case, the natural world isn't really a fan of hard and fast rules, as shown by the example of moss mites which have been quietly reproducing asexually since fish crawled out of the water and took their first tentative steps on land.

Moss mites (order Oribatida) are tiny arthropods that belong to the same group as spiders and look somewhat like a plump, shiny tick. There are about 12,000 described species (although some estimate that there could be ten times as many) found in the soil all around the globe. In fact, moss mites are thought to be some of the most prevalent arthropods found in forest soils, where they play a prominent role in breaking down organic matter, while at the same time helping to spread fungi around.

The mites are divided into many different groups, but what is interesting about these arthropods is that parthenogenesis – asexual reproduction in which a female only produces other female offspring (*see* New Mexico whiptail lizard) – has evolved independently multiple times and in many different branches of the mites' evolutionary tree. Because of this, scientists study moss mites to investigate why sex is so prevalent

Parthenogenesis has evolved multiple times in oribatid mites, with some lineages thought to have been reproducing this way for at least 400 million years.

in plants and animals and, furthermore, why some groups have given up on it entirely. To everyone's surprise, the studies revealed that parthenogenesis in moss mites is not a recent innovation: some lineages have been doing it for up to 400 million years. This extraordinary timeframe for an asexual organism to survive is more than ten times longer than the previous record, which itself had radically changed the perception of how parthenogenesis can persist.

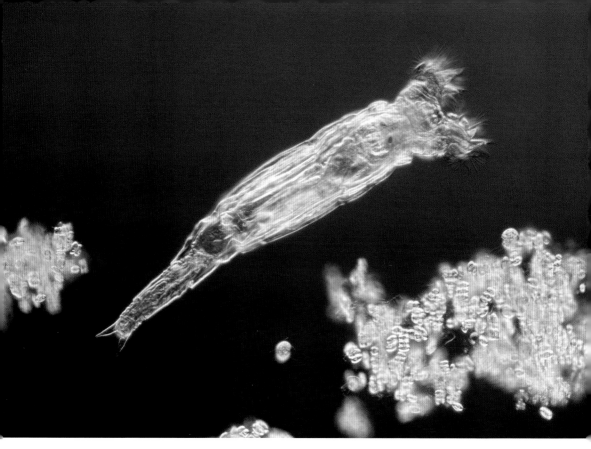

When scientists discovered that bdelloid rotifers had been reproducing asexually for 40 million years it radically changed how they thought about reproduction.

It is now thought that the mites become asexual species when they are living in incredibly food rich environments and extremely high population densities. These large populations in turn help to maintain a higher degree of genetic diversity and so allow the mites to shed any potentially damaging genetic mutations. Coupled with the fact that the environments in which they often live – boreal and temperate forest floors and peat bogs – are remarkably stable and have changed little since the Devonian Period some 400 million years ago, this has meant that the mites have had little need to change their ways.

But not only have the mites been in existence without having sex for the longest time in recorded history, they've also been splitting and creating new species. Considering how long they've been around this might come as no surprise, but this is actually the only parthenogenic animal species we know of that has diversified into multiple new ones. Typically a group of animals have to become reproductively isolated for this to happen, but if all the mites are reproducing on their own this becomes a little trickier. And moss mites are not alone in sustaining asexual reproduction over long periods of time.

Rotifers are a little-known group of miniature animals that live in freshwater environments, in moist soil, on bark and within moss. They are microscopic, but still classed as multicellular animals with specialized organs and a digestive tract that contains a mouth and an anus. Most of the 2,000 species of these little creatures are facultative parthenogens, which means that they can reproduce sexually and asexually, but a group known as the bdelloid rotifers have also done away with sex bit completely.

When scientists realized that the 450 or so species of bdelloid rotifers had been practising asexual reproduction for an impressive 40 million years, it radically changed what scientists thought about sex. The rotifers became known as the 'ancient asexuals', although perhaps the title should now be conferred on moss mites. The bdelloid rotifers and moss mites have helped scientists question their assumptions about sex, reproduction and the fundamentals of evolution in more ways than one. This is because in addition to maintaining asexuality for millions of years, the moss mites are also a rare example of parthenogenic species re-evolving sex.

This seems to contradict what in evolutionary biology is known as 'Dollo's law', which in general terms posits that once a complex trait is lost it is very unlikely to be re-evolved. As sex is presumably a 'complex trait', having probably evolved once very early during the evolution of life, theoretically it should be exceedingly difficult for it to have re-evolved again. But a few species, such as moss mites, plants in the genus *Hieracium*, crustaceans known as ostracods, have. Despite sex often being viewed as a fundamental for life on Earth, there is still a lot we don't know about it and scientists are constantly learning more.

Dungowan bush tomato
Changeable sex

In the early 1970s, botanists working in the Australian outback came across a curious plant. Growing up to 80 centimetres (31 inches) tall on a single, prickly stem, it had grey-green leaves and lovely, delicate purple flowers with a starkly contrasting yellow centre that matured into an almost perfectly round, green fruit. It clearly belonged to the nightshade plant family, the Solanaceae, which contains some 3,000–4,000 species in a huge range of sizes and shapes, from vines and epiphytes to shrubs and trees, and includes many familiar plants such as the tomato, potato, aubergine, bell pepper, chilli and tobacco. Found on every continent except Antarctica, and with around 90 new species described every year, it was no surprise to find another growing in the semi-arid forests of Australia. But this new plant appeared to defy further classification, because every plant specimen the botanists discovered seemed to show a different form of sexual expression.

Most species of nightshade produce what are known as 'cosexual' flowers (sometimes called hermaphrodite, bisexual or perfect flowers), which contain both the pollen-producing stamens and ovule-producing carpels. In loose terms, it means that each individual flower has both the 'male' and 'female' structures. But researchers were confused by the diversity of sexual expression in this new plant, leading them to think it might be a hybrid plant. In 2016, during fieldwork in the Northern Territory, Australia, researchers discovered a new population of the plants alongside a section of highway. Collecting new specimens, and returning a few years later to snaffle some more, they were able to gather a number of viable seeds, which they grew to plants in controlled conditions and compared with the 1970s specimens. From this they were finally able to understand what had so confused the earlier botanists – the bush tomato was doing something truly unusual, exhibiting three different breeding systems all at the same time.

Usually a plant falls into one main system of sexual reproduction. Roughly 85% of the around 250,000 species of flowering plants known to science (including most nightshades) exhibit the cosexual breeding system and can technically self-fertilize, although many have also developed ways to prevent this (*see* Chinese shell ginger).

The plants are unusual in showing three different types of reproductive behaviour, developing either only male flowers, cosexual flowers or both male and cosexual flowers.

The sexual reproduction of the Dungowan bush tomato confused botanists for almost half a century before it was finally understood what was happening.

The remaining 15% of flowering plants vary widely in how they reproduce. Some produce separate flowers that have 'male' and 'female' structures on the same plant, whilst others produce one or the other of the two flowers on individual plants (*see* European yew). Even further gradations exist, ranging from fully 'male' trees to fully 'female' trees and everything in between (*see* Common ash).

The bush tomato, however, blends a number of these different systems. Any given plant can produce only flowers with the male structures, known as 'functional dioecy', only cosexual flowers that contain both the male and female structures, or both cosexual and male flowers, which is known as 'andromonoecy'. This is highly unusual and, after genetic analysis and studying the subtle differences in leaf shape and a few other variations, the researchers were confident that the bush tomato represented an entirely new species. They named the plant *Solanum plastisexum*, from the Greek for 'changeable sex'. Why this bush tomato has evolved in such a way is not really known. But it does highlight the extraordinary flexibility of plants and their bodies, while demonstrating why using similar terminology to compare plants to animals can be complicated and clunky. When it comes to reproduction plants really do push the boundaries of the natural world if you take animals as the norm and try to fit everything else in around them.

Barklice
Sex-reversed genitals

The genitals of animals come in all shapes and sizes, with spines, hooks, dead ends and even jaws. While typically the male inserts an external appendage into the female to copulate, there are exceptions. In some mites and beetles for example, the females use a protruding organ to collect the males' sperm, while female seahorses use an ovipositor to insert eggs into the male's pouch. But few genitals in the animal world turn the tables quite as dramatically as barklice in the genus *Neotrogla*.

Female barklice have, extraordinarily, evolved a 'penis' that they insert into the male 'vagina' to extract packages of sperm and nutrients. The penis fully inflates when inside the male's cavity, which has perfectly evolved to fit the female's appendage. The penis-like structure then latches on with spines to prevent the male from escaping, before the male ejaculates within its own body cavity and the sperm is transfered to the female along ducts on the appendage. The organ is, more accurately, described as a 'gynosome', but it achieves all the things that the male penis does – it transfers male sex cells from one individual to the next. The only real difference is that instead of delivering sperm, the gynosome collects it. This is the first animal discovered with 'sex-reversed' genitalia. But what is more incredible still is that the gynosome has evolved twice.

Barklice belong to a group of insects known as the Psocoptera, roughly 6,000 species ranging in size from 1 to 10 millimetres (up to ½ inch) long. These typically tiny insects have antennae and two sets of paired wings, although some species have evolved to become wingless. The insects primarily feed on fungi, algae and lichen growing on tree bark, giving rise to their more common name. But the four species in the genus *Neotrogla* have given up their forest-living ways and opted for a subterranean life, living in dry cave systems throughout Brazil where they feed on bat guano. But the *Neotrogla* barklice are not alone in evolving a gynosome. Research has found that females in another group of barklice, the *Afrotrogla,* also have a 'penis'. The last common ancestor of the two genera have a more traditional male penis and female vagina, so it seems that the *Neotrogla* and *Afrotrogla* insects have evolved the structure independently. Intriguingly, both groups of

Barklice in the genus *Neotrogla* are unusual in the animal world in that the females have effectively evolved a 'penis' with which they penetrate the males to retrieve their sperm.

barklice live in caves, leading scientists to hypothesize that the shift to a nutrient-poor environment may have driven the evolution of the gynosome.

It all comes down to resources. Typically, males spend fewer resources on creating each sperm cell, allowing them to produce far more sex cells than females, which in turn produce fewer, more resource-intensive egg cells. Because females don't want to waste their few, precious eggs, they tend to be more discerning about who they mate with. As a result, males have to compete for females, either through fighting or showing off their fancy looks with a dance. In the nutrient-poor cave systems, the female barklice need as much food as possible. As a result, the males package their sperm up with nutrients, giving the female a meal when they mate. So, rather than males competing for the limited number of eggs produced by females, the females compete for the limited number of resource-heavy sperm and nutrient packages produced by the males. This may have resulted in the evolution of the gynosome. It not only allows the females to channel the sperm packages into their bodies, but a special switching valve at the entrance of the organ storing semen means that the female can store multiple packages, intensifying the competition between females for males and further driving the structures' evolution.

The gynosomes of *Neotrogla* and *Afrotrogla* are extraordinary examples of convergent evolution – when organisms living in similar environments or undertaking similar tasks evolve strikingly similar structures. These two groups of insects are the only known animals in which the female penetrates the male during sex. The discovery of this structural reversal in *Neotrogla* and *Afrotrogla* barklice has made scientists reassess exactly what it means to be male and what it means to be female.

Index

Further reading

Adriaens, P.R. and De Block, A., 2022, *Of Maybugs and Men: A History and Philosophy of the Sciences of Homosexuality*, University of Chicago Press. **Bagemihl, B.**, 1999, *Biological Exuberance: Animal Homosexuality and Natural Diversity*, Stonewall Inn Editions. **Brooks, R.**, 2021, Darwin's closet: the queer sides of *The descent of man* (1871*). Zoological Journal of the Linnean Society*, 191(2): 323-346. **McLaughlin, J.F. et al.**, 2023, Multivariate models of animal sex: breaking binaries leads to a better understanding of ecology and evolution. *Integrative and Comparative Biology*, https://doi.org/10.1093/icb/icad027. **Monk, J.D. et al.**,

2019, An alternative hypothesis for the evolution of same-sex sexual behaviour in animals. *Nature Ecology & Evolution*, 3: 1622-1631.**Roughgarden, J.**, 2009, *Evolution's Rainbow: Diversity, Gender, and Sexuality in Nature and People*, University of California Press, 2nd edn. **Sandford, S.**, 2022, *Vegetal Sex: Philosophy of Plants*, Bloomsbury Academic. **Schrefer, E.**, 2022, *Queer Ducks (and Other Animals): The Natural World of Animal Sexuality*, Katherine Tegen Books. **Sommer, V. and Vasey, P.L.**, 2011, *Homosexual Behaviour in Animals: An Evolutionary Perspective*, Cambridge University Press.

References

All websites correct at time of printing.

Introduction

Bagemihl, B. (1999), *Biological Exuberance: Animal Homosexuality and Natural Diversity*, Stonewall Inn Editions **Clay, Z. and de Waal, F.B.M. (2015)**, Sex and strife: post-conflict sexual contacts in bonobos: https://brill.com/view/journals/beh/152/3-4/article-p313_4.xml **Clay, Z. and Zuberbühler, K. (2012)**, Communication during sex among female bonobos: effects of dominance, solicitation and audience: https://www.nature.com/articles/srep00291 **Gómez, J.M. et al. (2023)**, The evolution of same-sex sexual behaviour in mammals: https://www.nature.com/articles/s41467-023-41290-x **IUCN Red List (2021)**: https://www.iucnredlist.org/resources/summary-statistics **Love, T.M. (2019)**: The impact of oxytocin on stress: the role of sex: https://www.ncbi.nlm.nih.gov/pmc/articles/PMC6863168/ **McLaughlin, J.F. et al. (2023)**, Multivariate Models of Animal Sex: Breaking Binaries Leads to a Better Understanding of Ecology and Evolution: https://academic.oup.com/icb/article/63/4/891/7157109 **Troisi, A. and Carosi, M. (1998)**, Female orgasm rate increases with male dominance in Japanese macaques: https://www.sciencedirect.com/science/article/abs/pii/S0003347298908983

Adélie penguin

Davis, J. (2020), Terra Nova notebooks describing penguin sexual behaviours acquired by the Museum: https://www.nhm.ac.uk/discover/news/2020/april/terra-nova-notebooks-penguin-sexual-behaviours-acquired-by-museum.html **Gao, Y. et al. (2022)**, The occupation history of the longest-dwelling Adélie penguin colony reflects Holocene climatic and environmental changes in the Ross Sea, Antarctica: https://www.sciencedirect.com/science/article/pii/S0277379122001251. **Pincemy, G. et al (2010)**, Homosexual Mating Displays in Penguins: https://onlinelibrary.wiley.com/doi/full/10.1111/j.1439-0310.2010.01835.x **Richardson, J. and Parnell, P. (2005)**, *And Tango Makes Three*, Simon and Schuster Children's UK **Robson, D. (2011)**, The Scott expedition: how science gained the pole position: https://www.telegraph.co.uk/news/science/science-news/8587530/The-Scott-expedition-how-science-gained-the-pole-position.html **Russell, D.G.D. et al. (2012)**, Dr. George Murray Levick (1876–1956): unpublished notes on the sexual habits of the Adélie penguin: https://www.cambridge.org/core/journals/polar-record/article/abs/dr-george-murray-levick-18761956-unpublished-notes-on-the-sexual-habits-of-the-adelie-penguin/8647660D29AD9660C9C16623638C9116

Mangrove killifish

Edenbrow, M. (2011), Behavioural types and life history strategies during ontogeny in the mangrove killifish, *Kryptolebias marmoratus*: https://linkinghub.elsevier.com/retrieve/pii/S0003347211002831 **Kelley, J.L. et al. (2016)**, The Genome of the Self-Fertilizing Mangrove Rivulus Fish, Kryptolebias marmoratus: A Model for Studying Phenotypic Plasticity and Adaptations to Extreme Environments: https://www.ncbi.nlm.nih.gov/pmc/articles/PMC4987111/ **Lubinski, B.A. et al. (1995)**, Outcrossing in a Natural Population of a Self-Fertilizing Hermaphroditic Fish: https://academic.oup.com/jhered/article-abstract/86/6/469/801320?redirectedFrom=fulltext&login=false **Sakakura, Y. et al. (2006)**, Gonadal morphology in the self-fertilizing mangrove killifish, *Kryptolebias marmoratus*: https://link.springer.com/article/10.1007/s10228-006-0362-2 **Taylor, D.S. (2012)**, Twenty-Four Years in the Mud: What Have We Learned About the Natural History and Ecology of the Mangrove Rivulus, *Kryptolebias marmoratus*?: https://www.ncbi.nlm.nih.gov/pmc/articles/PMC3501094

Duck-billed dinosaur

Cates, K. (2014), Bynum dinosaur pioneer Marion Brandvold dies at 102: https://eu.greatfallstribune.com/story/news/local/2014/06/09/choteau-dinosaur-pioneer-marion-brandvold-dies/10261091/ **Davis, J. (2020)**, How to sex a dinosaur: https://www.nhm.ac.uk/discover/how-to-sex-a-dinosaur.html **Horner, J. and Makela, R. (1979)**, Nest of juveniles provides evidence of family structure among dinosaurs: https://www.nature.com/articles/282296a0 **Knapp, A. et al. (2021)**, Three-dimensional geometric morphometric analysis of the skull of Protoceratops andrewsi supports a socio-sexual signalling role for the ceratopsian frill: https://royalsocietypublishing.org/doi/10.1098/rspb.2020.2938 **Maiorino,**

L. et al. (2015), Males Resemble Females: Re-Evaluating Sexual Dimorphism in *Protoceratops andrewsi* (Neoceratopsia, Protoceratopsidae): https://journals.plos.org/plosone/article?id=10.1371/journal.pone.0126464 **Peters, W.S. and Peters, D.S. (2009)**, Life history, sexual dimorphism and 'ornamental' feathers in the mesozoic bird *Confuciusornis sanctus*: https://www.ncbi.nlm.nih.gov/pmc/articles/PMC2828012/ **Weishampel, D.B. and Reif, W-E. (1984)**, The Work of Franz Baron Nopcsa (1877–1933): Dinosaurs, Evolution and Theoretical Tectonics: https://www.zobodat.at/pdf/JbGeolReichsanst_127_0187.pdf

New Mexico whiptail lizard

Cowan, D.P. and Stahlhut, J.K. (2004), Functionally reproductive diploid and haploid males in an inbreeding hymenopteran with complementary sex determination): https://www.ncbi.nlm.nih.gov/pmc/articles/PMC478579/ **Crews, D. et al. (1986)**, Behavioral facilitation of reproduction in sexual and unisexual whiptail lizards: https://www.pnas.org/doi/10.1073/pnas.83.24.9547 **Dubey, S. et al. (2019)**, Population genomics of an exceptional hybridogenetic system of *Pelophylax* water frogs: https://bmcecolevol.biomedcentral.com/articles/10.1186/s12862-019-1482-4 **Harmon, K. (2010)**, No Sex Needed: All-Female Lizard Species Cross Their Chromosomes to Make Babies: https://www.scientificamerican.com/article/asexual-lizards/ **Hegarty, J.M. and Hiscock, S.J. (2004)**, Hybrid speciation in plants: new insights from molecular studies: https://nph.onlinelibrary.wiley.com/doi/10.1111/j.1469-8137.2004.01253.x **Larsen, P.A. et al. (2010)**, Natural hybridization generates mammalian lineage with species characteristics: https://www.ncbi.nlm.nih.gov/pmc/articles/PMC2895066/ **Liegeois M. et al. (2021)**, Extremely Widespread Parthenogenesis and a Trade-Off Between Alternative Forms of Reproduction in Mayflies (Ephemeroptera): https://academic.oup.com/jhered/article/112/1/45/5904753 **Lues, A.A. et al. (2011)**, Laboratory synthesis of an independently reproducing vertebrate species: https://www.pnas.org/doi/full/10.1073/pnas.1102811108 **Lutes, A.A. et al. (2010)**, Sister Chromosome Pairing Maintains Heterozygosity in Parthenogenetic Lizards: https://www.ncbi.nlm.nih.gov/pmc/articles/PMC2840635/ **Mallet, J. (2007)**, Hybrid speciation: https://www.nature.com/articles/nature05706?report=reader **Mata-Silva, V. et al. (2010)**, Reproductive characteristics of two syntopic whiptail lizards, *Aspidoscelis marmorata* and *Aspidoscelis tesselata*, from the northern Chihuahuan Desert: https://www-jstor-org.ezproxy.nhm.ac.uk/stable/40588616?sid=primo#metadata_info_tab_contents **Mavárez, J. et al. (2006)**, Speciation by hybridization in *Heliconius* butterflies: https://www.nature.com/articles/nature04738 **Morales-Hojas, R. et al. (2020)**, Population genetic structure and predominance of cyclical parthenogenesis in the bird cherry-oat aphid *Rhopalosiphum padi*: https://www.ncbi.nlm.nih.gov/pmc/articles/PMC7232763/ **Neaves, W.B. and Baumann, P. (2011)**, Unisexual reproduction among vertebrates: https://www.sciencedirect.com/science/article/abs/pii/S0168952510002295 **Watts, P.C. et al. (2006)**, Parthenogenesis in Komodo dragons: https://www.nature.com/articles/4441021a **Yam, P. (2006)**, Strange but True: Komodo Dragons Show that 'Virgin Births' Are Possible: https://www.scientificamerican.com/article/strange-but-true-komodo-d/ **Yong, E. (2010)**, Extra chromosomes allow all-female lizards to reproduce without males: https://www.nationalgeographic.com/science/article/extra-chromosomes-allow-all-female-lizards-to-reproduce-without-males

Morpho butterfly

Asdell, S.A. (1942), The Accessory Reproductive Tract in Mammalian True Hermaphrodites, an Effect of Position: https://www.jstor.org/stable/2457667 **Broad, G. and Shaw, R. (2018)**, Two examples of anterior/posterior gynandromorphism in Ichneumonidae (hymenoptera): http://www.filming-varwild.com/articles/mark_shaw/318_Gynandromorph_Ichs.pdf **Hollander, W.F. et al. (1956)**, A study of 25 gynandromorphic mice of the bagg albino strain: https://onlinelibrary.wiley.com/doi/10.1002/ar.1091240207 **Krohmer, R.W. (1989)**, Reproductive Physiology and Behavior of a Gynandromorph Redsided Garter Snake, T.*hamnophis sirtalis parietalis*, from Central Manitoba, Canada: https://www.jstor.org/stable/1446001 **Krumm, J.L. (2013)**, Axial Gynandromorphy and Sex Determination in *Branchinecta lindahli* (Branchiopoda: Anostraca): https://academic.oup.com/jcb/article/33/3/303/2548114 **Mitchell, J.C. and Fouquette, Jr., M.J. (1978)**, A Gynandromorphic Whiptail Lizard, *Cnemidophorus inornatus*, from Arizonan England: https://www.jstor.org/stable/1443840 **Olmstead, A.W. and LeBlanc, G.A. (2006)**, The Environmental-Endocrine Basis of Gynandromorphism (Intersex) in a Crustacean: https://www.ncbi.nlm.nih.gov/pmc/articles/PMC1752225/ **Palmgren, P. (1979)**, On the frequency of gynandromorphic spiders: https://www.jstor.org/stable/23733636 **Pavid, K. (2015)**, Beauty of the dual-gender butterfly: https://www.nhm.ac.uk/discover/beauty-dual-gender-butterfly.html **Weintraub, K. (2019)**, Split-sex animals are unusual, yes, but not as rare as you'd think: https://www.nytimes.com/2019/02/25/science/split-sex-gynandromorph.html **Werner, E. (2012)**, A Developmental Network Theory of Gynandromorphs, Sexual Dimorphism and Species Formation: https://www.researchgate.net/publication/233967450_A_Developmental_Network_Theory_of_Gynandromorphs_Sexual_Dimorphism_andSpecies_Formation

Western lowland gorilla

Bagemihl, B. (1999), *Biological Exuberance: Animal Homosexuality and Natural Diversity*, Stonewall Inn Editions. **Grueter, C.C. and Stoinski, T.S. (2016)**, Homosexual Behavior in Female Mountain Gorillas: Reflection of Dominance, Affiliation, Reconciliation or Arousal?: https://journals.plos.org/plosone/article?id=10.1371/journal.pone.0154185#pone.0154185.ref018 **de Waal, F. (2022)**, *Different: what apes can teach us about gender*, Granta Books. **Yamagiwa, J. (1987)**, Intra- and inter-group interactions of an all-male group of Virunga mountain gorillas: https://link.springer.com/article/10.1007/BF02382180

Domestic sheep

Bagemihl, B. (1999), *Biological Exuberance: Animal Homosexuality and Natural Diversity*, Stonewall Inn Editions **Bhattacharya, S. et al. (2022)**, Identification of differential hypothalamic DNA methylation and gene expression associated with sexual partner preferences in rams: https://journals.plos.org/plosone/article?id=10.1371/journal.pone.0263319 **Hegde, N.G. (2019)**, Livestock development for sustainable livelihood of small farmers: https://www.researchgate.net/publication/332947789_Livestock_Development_for_Sustainable_Livelihood_of_Small_Farmers **Meadows, J.R.S. et al. (2007)**, Five Ovine Mitochondrial Lineages Identified From Sheep Breeds of the Near East: https://www.ncbi.nlm.nih.gov/pmc/articles/PMC1840082/ **Perkins, A. and Roselli, C.E. (2008)**, The Ram as a Model for Behavioral Neuroendocrinology: https://www.ncbi.nlm.nih.gov/pmc/articles/PMC2150593/ **Roselli, C.E. et al. (2004)**, Sexual partner preference, hypothalamic morphology and aromatase in rams: https://pubmed.ncbi.nlm.nih.gov/15488542/ **Schrefer, E. (2022)**, Queer animals are everywhere. Science is finally catching on: https://www.washingtonpost.com/magazine/2022/06/30/queer-animals-are-everywhere-science-is-finally-catching/ **Schwartz, J. (2007)**, Of gay sheep, modern science and bad publicity: https://www.nytimes.com/2007/01/25/science/25sheep.html **Vasey, P. (2002)**, Sexual partner preference in female Japanese macaques: https://pubmed.ncbi.nlm.nih.gov/11910792/ **Vasey, P. et al. (2014)**, Female homosexual behavior and inter-sexual mate competition in Japanese macaques: Possible implications for sexual selection theory: https://www.sciencedirect.com/science/article/abs/pii/S0149763414002176

Saharan cypress

Foucaud, J. et al. (2010), Thelytokous parthenogenesis, male clonality and genetic caste determination in the little fire ant: new evidence and insights from the lab: https://www.nature.com/articles/hdy2009169 **Lenin, J. (2017)**, The Male Fish That Fathered a Clone of Himself: https://thewire.in/environment/androgenesis-alburnoides-cloning-sexual-vertebrate **Mantovani, B. and Scali, V. (1992)**, Hybridogenesis and androgenesis in the stick-insect Badcillus rossius-grandii benazzzii (Insecta, Phasmatodea): https://pubmed.ncbi.nlm.nih.gov/28568678/ **Nava, J.L.R. et al. (2010)**, Molecular evidence for the natural production of homozygous *Cupressus sempervirens* L. lines by Cupressus dupreziana seed trees: https://www.nature.com/articles/hdy2009112 **Pichot, C. (2001)**, Surrogate mother for endangered *Cupressus*: https://www.nature.com/articles/35083687 **Pichot, C. (2000)**, Unreduced diploid nuclei in *Cupressus dupreziana* A. Camus pollen: https://link.springer.com/article/10.1007/s001220051518 **Pichot, C. et al. (2000)**, Lack of mother tree alleles in zymograms of *Cupressus dupreziana* A. Camus embryos: https://www.afs-journal.org/articles/forest/abs/2000/01/asf0103/asf0103.html **Werner, L. (2007)**, A cypress in the Sahara: https://archive.aramcoworld.com/issue/200705/a.cypress.in.the.sahara.htm

Bicolour parrotfish

Barba, J. (2010), Demography of parrotfish: age, size and reproductive variables: https://researchonline.jcu.edu.au/26682/ **Brook, H.J. et al. (1994)**, Protogynous Sex Change in the Intertidal Isopod *Gnorimosphaeroma oregonense* (Crustacea: Isopoda): https://pubmed.ncbi.nlm.nih.gov/29281308/ **Casas, L. et al. (2016)**, Sex Change in Clownfish: Molecular Insights from Transcriptome Analysis: https://www.nature.com/articles/srep35461 **Christenhusz, M.J.M. and Byng, J.W. (2016)**, The number of known plants species in the world and its annual increase: https://www.biotaxa.org/Phytotaxa/article/view/phytotaxa.261.3.1 **Fishelson, L. and Galil, B.S. (2001)**, Gonad Structure and Reproductive Cycle in the Deep-Sea Hermaphrodite Tripodfish, *Bathypterois mediterraneus* (Chlorophthalmidae, Teleostei): https://www.jstor.org/stable/1447905 **Gemmell, N.J. et al. (2019)**; Chapter Three – Natural sex change in fish: https://www.sciencedirect.com/science/article/abs/pii/S0070215318301145 **He, Z. et al. (2022)**, Crosstalk between sex-related genes and apoptosis signaling reveals molecular insights into sex change in a protogynous hermaphroditic teleost fish, ricefield eel *Monopterus albus*: https://www.sciencedirect.com/science/article/abs/pii/S0044848622000321 **Jarne, P. and Auld, J.R. (2006)**, Animals mix it up too: The distribution of self-fertilisation among hermaphroditic animals: https://bioone.org/journals/evolution/volume-60/issue-9/06-246.1/ANIMALS-MIX-IT-UP-TOO--THE-DISTRIBUTION-OF-SELF/10.1554/06-246.1.short **Osterloff, E. (2023)**, The unusual link between parrotfish and sand: https://www.nhm.ac.uk/discover/parrotfish-and-sand.html **Pavlowich, T. et al. (2018)**, Leveraging sex change in parrotfish to manage fished populations: https://www.researchgate.net/publication/328152945_Leveraging_sex_change_in_parrotfish_to_manage_fished_populations **Rodgers, E.W. et al. (2007)**, Social status determines sexual phenotype in the bi-directional sex changing bluebanded goby *Lythrypnus dalli*: https://onlinelibrary.wiley.com/doi/10.1111/j.1095-8649.2007.01427.x **Roughgarden, J. (2004)**, *Evolution's Rainbow: Diversity, Gender, and Sexuality in Nature and People*, University of California Press **Royer, M. (1975)**, Hermaphroditism in Insects. Studies on *Icerya purchase*: https://link.springer.com/chapter/10.1007/978-3-642-66069-6_14 **Russell-Hunter, W.D. and McMahon, R.F. (1976)**, Evidence for Functional Protandry in a Fresh-Water Basommatophoran Limpet, *Laevapex fuscus*: https://www.jstor.org/stable/3225061

Swans

Bagemihl, B. (1999), *Biological Exuberance: Animal Homosexuality and Natural Diversity*, Stonewall Inn Editions. **BirdLife International (2023)**, Species factsheet: *Cygnus atratus*: http://datazone.birdlife.org/species/factsheet/black-swan-cygnus-atratus/distribution **Braithwaite, L.W. (1981)**, Ecological Studies of the Black Swan III. Behaviour and Social Organisation, Australian Wildlife Research, 8(1): 135–146: https://www.publish.csiro.au/wr/WR9810135 **Brugger, C. and Taborsky, M. (1994)**, Male Incubation and its Effect on Reproductive Success in the Black Swan, *Cygnus atratus*: *Ethology* 96(2): 138-146 https://onlinelibrary.wiley.com/doi/abs/10.1111/j.1439-0310.1994.tb00889.x **Carboneras, C. and G.M. Kirwan (2020)**, Black Swan (*Cygnus atratus*), version 1.0. In *Birds of the World* (J. del Hoyo, A. Elliott, J. Sargatal, D.A. Christie, and E. de Juana, Editors), Cornell Lab of Ornithology, Ithaca, NY, USA: https://doi.org/10.2173/bow.blkswa.01 **Kraaijeveld, K. et al. (2004)**, Extra-pair paternity does not result in differential sexual selection in the mutually ornamented black swan (*Cygnus atratus*): https://pubmed.ncbi.nlm.nih.gov/15140105/ **Ritchie, J. (1926)**, Nesting of two male swans, *The Scottish Naturalist* 157: 95–96: https://www.biodiversitylibrary.org/item/204086#page/114/mode/1up

Green sea turtle

Blechschmidt, J. et al. (2020); Climate Change and Green Sea Turtle Sex Ratio - Preventing Possible Extinction: https://www.ncbi.nlm.nih.gov/pmc/articles/PMC7288305/ **Bock, S.L. et al. (2020)**, Spatial and temporal variation in nest temperatures forecasts sex ratio skews in a crocodilian with environmental sex determination: https://pubmed.ncbi.nlm.nih.gov/32345164/ **Dissanayake, D.S.B. et al. (2021)**, Effects of natural nest temperatures on sex reversal and sex ratios in an Australian alpine skink: https://www.nature.com/articles/s41598-021-99702-1 **González, E.J. et al. (2019)**, The sex-determination pattern in crocodilians: A systematic review of three decades of research:

https://besjournals.onlinelibrary.wiley.com/doi/full/10.1111/1365-2656.13037 **Hannam, P. (2014)**, Hotter climate could turn sea turtles all-girl: https://www.smh.com.au/environment/conservation/hotter-climate-could-turn-sea-turtles-allgirl-20140518-38hy2.html **Jones, M. et al. (2020)**, Reproductive phenotype predicts adult bite-force performance in sex-reversed dragons (*Pogona vitticeps*): https://onlinelibrary.wiley.com/doi/full/10.1002/jez.2353 **Moeller, K.T. (2013)**, Temperature-Dependent Sex Determination in Reptiles: https://embryo.asu.edu/pages/temperature-dependent-sex-determination-reptiles **NOAA (2023)**, What causes a sea turtle to be born male or female?: https://oceanservice.noaa.gov/facts/temperature-dependent.html **Woodward, D.E. and Dickson Murray, J. (1993)**, On the effect of temperature-dependent sex determination on sex ratio and survivorship in crocodilians: https://royalsocietypublishing.org/doi/10.1098/rspb.1993.0059

Giraffe

Bagemih, B. (1999), *Biological Exuberance: Animal Homosexuality and Natural Diversity*, Stonewall Inn Editions **Hymas, C. (2019)**, Labour embroiled in bizarre row over whether giraffes are gay: https://www.telegraph.co.uk/politics/2019/10/27/labour-embroiled-bizarre-row-whether-giraffes-gay/ **Innis Dagg, A. (1984)**, Homosexual behaviour and female-male mounting in mammals - a first survey: 1984; https://onlinelibrary.wiley.com/doi/abs/10.1111/j.1365-2907.1984.tb00344.x?casa_token=6lo2GgLKcboAAAAA:sCpUFfk-9nR1vVEhWWDTIWPOzDhs4ESz9LZls8Vg8YBgv1xpX_B937czgfM63iCsaAVa8SNDYk8H **MacLellan, L. (2020)**, 'Tenure Not Granted': A new film honors a woman science forgot: https://qz.com/work/1782647/giraffe-researcher-anne-innis-dagg-a-woman-science-forgot **Milton, J. (2020)**, Dawn Butler on her Labour deputy leader bid, that 'gay giraffe' scandal and her fearless commitment to trans rights: https://www.thepinknews.com/2020/01/13/dawn-butler-labour-party-mp-deputy-leader-bid-gay-giraffes-gra-trans-rights/ **Mitchell, A. (2019)**, The curious, extraordinary life of Anne Innis Dagg: https://canadiangeographic.ca/articles/the-curious-extraordinary-life-of-anne-innis-dagg/ **Ondrako, S. (2021)**, Giraffes & Feminism – Dr. Anne Innis Dagg: https://www.youtube.com/watch?v=28HQvLl6yjw&ab_channel=TheSaraOndrakoShow **Owen, G. (2019)**, Gay giraffe row splits upper ranks of Labour party over MP Dawn Butler is ridiculed by senior advisers for suggesting 99% of the animals are sexually attracted to their own gender: https://www.dailymail.co.uk/news/article-7617913/Gay-giraffe-row-splits-upper-ranks-Labour-party-Dawn-Butler-ridiculed-senior-advisers.html **Peters, D. (2019)**, Pioneering biologist Anne Innis Dagg gets her due: https://www.universityaffairs.ca/features/feature-article/pioneering-biologist-anne-innis-dagg-gets-her-due/

Common ash

Ainsworth, C. (2000), Boys and Girls Come Out to Play: The Molecular Biology of Dioecious Plants: https://academic.oup.com/aob/article/86/2/211/2588220 **Albert, B. et al. (2013)**, Sex expression and reproductive biology in a tree species, *Fraxinus excelsior* L: https://www.sciencedirect.com/science/article/pii/S1631069113001923 **Bauer, R.T. (1986)**, Sex change and life history pattern in the shrimp *Thor manningi* (Decopoda: Caridea): A novel case of partial protandric hermaphroditism: https://decapoda.nhm.org/pdfs/31531/31531.pdf **Godin, V.N. (2023)**, Trioecy in Flowering Plants: https://link.springer.com/article/10.1134/S0012496622060023 **Golding, B. (1996)**, Evolution: When was life's first branch point?: https://www.sciencedirect.com/science/article/pii/S0960982209004485 **Joppa, L.N. et al. (2010)**; How many species of flowering plants are there?: https://royalsocietypublishing.org/doi/10.1098/rspb.2010.1004 **Oyarzún, P.A. et al. (2020)**, Trioecy in the Marine Mussel *Semimytilus algosus* (Mollusca, Bivalvia): Stable Sex Ratios Across 22 Degrees of a Latitudinal Gradient: https://www.frontiersin.org/articles/10.3389/fmars.2020.00348/full **Perera, P.I.P. et al. (2010)**, Early inflorescence and floral development in Cocos nucifera L. (Arecaceae: Arecoideae): https://www.sciencedirect.com/science/article/pii/S0254629910001195 **Schlesinger, A. et al. (2010)**, Sexual Plasticity and Self-Fertilization in the Sea Anemone Aiptasia diaphana: https://www.ncbi.nlm.nih.gov/pmc/articles/PMC2912375/ **Sohn, J.J. and Policansky, D. (1977)**, The Costs of Reproduction in the Mayapple *Podophyllum peltatum* (Berberidaceae): https://esajournals.onlinelibrary.wiley.com/doi/abs/10.2307/1935088

Common cockchafer

Adriaens, P. and De Block, A. (2022), *Of Maybugs and Men*, University of Chicago Press **Brooks, R (2011)**, All too human: responses to same-sex copulation in the common cockchafer (*Melolontha melolontha* (L.)), 1834–1900: https://www.euppublishing.com/doi/full/10.3366/E0260954108000703 **Handbooks for the identification of British insects (2012)**: https://www.royensoc.co.uk/wp-content/uploads/2021/12/Vol05_Part11.pdf

European yew

BBC News (2015), Berries show ancient Fortingall yew tree is 'changing sex': https://www.bbc.co.uk/news/uk-scotland-tayside-central-34700033 **Blake-Mahmud, J. and Struwe, L. (2020)**, When the going gets tough, the tough turn female: injury and sex expression in a sex-changing tree: https://www.ncbi.nlm.nih.gov/pmc/articles/PMC7155049/ **Coleman, M (2015)**, Oldest yew tree switches sex: https://stories.rbge.org.uk/archives/17622 **Visit Scotland, Fortingall Yew**: https://www.visitscotland.com/info/see-do/fortingall-yew-p2568631 **Khanduri, V.P. et al. (2021)**, Gender plasticity uncovers multiple sexual morphs in natural populations of *Cedrus deodara* (Roxb.) G. Do: https://ecologicalprocesses.springeropen.com/articles/10.1186/s13717-021-00311-7 **Kite, G.C. (2013)**, Analysis of yew wood: https://www.kew.org/read-and-watch/analysis-of-yew-wood **Ne'eman, G. et al. (2011)**, Relationships between tree size, crown shape, gender segregation and sex allocation in *Pinus halepensis*, a Mediterranean pine tree: https://www.ncbi.nlm.nih.gov/pmc/articles/PMC3119615/ **Wheeler, L.C. and Mustoe, G. (1982)**, Pinicae: https://link.springer.com/referenceworkentry/10.1007/0-387-30843-1_329 **Wright, C. (2016)**, Can trees really change sex?: https://theconversation.com/can-trees-really-change-sex-50226

European eel

Colombo, G. and Grandi, G. (1996), Histological study of the development and sex differentiation of the gonad in the European eel: https://onlinelibrary.wiley.com/doi/abs/10.1111/j.1095-8649.1996.tb01443.x **Davey, A.J.H. and Jellyman, D.J. (2005)**, Sex Determination in Freshwater Eels and Management Options for Manipulation of Sex: https://link.springer.com/article/10.1007/s11160-005-7431-x **Geffroy, B. et al. (2013)**, New insights regarding gonad development in European eel: evidence for a direct ovarian differentiation: https://pubmed.ncbi.nlm.nih.gov/23334566/ **Geffroy, B., et al. (2016)**, Sexually dimorphic gene expressions in eels: useful markers for early sex assessment in a conservation context: https://www.ncbi.nlm.nih.gov/pmc/articles/PMC5034313/ **Lee, A. (2020)**, Sexual Eeling: The slippery subject of eel reproduction evaded human understanding for millennia: https://www.historytoday.com/archive/natural-histories/sexual-eeling

White-throated sparrow

Arnold, C. (2016), The sparrow with four sexes: https://www.nature.com/articles/539482a **Campagna, L. (2015)**, Supergenes: The Genomic Architecture of a Bird with Four Sexes: https://www.cell.com/current-biology/fulltext/S0960-9822(15)01484-0 **Grrl Scientist (2011)**, Sparrows show us a new way to have sexes: https://www.theguardian.com/science/punctuated-equilibrium/2011/may/25/2 **McLaughlin, J.F. et al. (2023)**, Multimodal models of animal sex: breaking binaries leads to a better understanding of ecology and evolution: https://www.biorxiv.org/content/10.1101/2023.01.26.525769v1.full.pdf **Tuttle, E.M. et al. (2016)**, Divergence and Functional Degradation of a Sex Chromosome-like Supergene: https://www.sciencedirect.com/science/article/pii/S0960982215015626?via%3Dihub

Spotted hyena

Cook, L. (2019), Everything you know about hyenas is wrong - these animals are fierce, social and incredibly smart: https://ideas.ted.com/everything-you-know-about-hyenas-is-wrong-these-animals-are-fierce-social-and-incredibly-smart/ **Cunha, G.R. et al. (2015)**, Development of the External Genitalia: Perspectives from the Spotted Hyena (*Crocuta crocuta*): https://www.ncbi.nlm.nih.gov/pmc/articles/PMC4069199/ **Cunha, G.R. et al. (2005)**, The Ontogeny of the Urogenital System of the Spotted Hyena (*Crocuta crocuta* Erxleben): https://academic.oup.com/biolreprod/article/73/3/554/2666898 **Funk, H. (2011)**, R. J. Gordon's Discovery of the Spotted Hyena's Extraordinary Genitalia in 1777:

https://link.springer.com/article/10.1007/s10739-011-9285-5 **Roughgarden, J. (2004),** *Evolution's Rainbow: Diversity, Gender, and Sexuality in Nature and People*, University of California Press **Wynn, R.M. and Amoroso, E.C. (1964),** Platentation in the spotted hyena (*Crocuta crocuta*), with particular reference to the circulation: https://onlinelibrary.wiley.com/doi/abs/10.1002/aja.1001150208

Western gull

Bagemihl, B. (2000), *Biological Exuberance: Animal Homosexuality and Natural Diversity*, Stonewall Inn Editions **Boxall, B. (1993),** The Nest Quest: Group Sets Sail for Lesbian Sea Gulls : Nature: Biologist takes gay group to Channel Islands, where he had observed same-sex pairs in the 1970s. Research has prompted controversy: https://www.latimes.com/archives/la-xpm-1993-06-20-me-5238-story.html **Conover, M.R. et al. (1979),** Female-Female Pairs and Other Unusual Reproductive Associations in Ring-Billed and California Gulls: https://www.jstor.org/stable/4085395?origin=JSTOR-pdf **New York Times (1977),** Extensive Homosexuality Is Found Among Seagulls Off Coast of California: https://www.nytimes.com/1977/11/23/archives/extensive-homosexuality-is-found-among-seagulls-off-coast-of.html **Hunt, G.L. and Hunt, M.W. (1977),** Female-Female Pairing in Western Gulls (*Larus occidentalis*) in Southern California: https://www.science.org/doi/10.1126/science.196.4297.1466 **Schlanger, Z. (2017),** The gulls are alright: How a lesbian seagull discovery shook up 1970s conservatives: https://qz.com/1023638/the-gulls-are-alright-how-a-lesbian-seagull-discovery-shook-up-1970s-conservatives **Young, L.C. et al. (2008),** Successful same-sex pairing in Laysan albatross: https://www.ncbi.nlm.nih.gov/pmc/articles/PMC2610150/

Common bottlenose dolphin

Ashworth, J. (2023), Calls for the UK to legally ban keeping whales and dolphins in captivity: https://www.nhm.ac.uk/discover/news/2023/september/calls-for-uk-ban-keeping-whales-dolphins-captivity.html **Acosta, N.B. (2015),** Same-Sex Socio-Sexual Interactions Among a Group of Captive Bottlenose Dolphins (Tursiops truncatus): https://aquila.usm.edu/cgi/viewcontent.cgi?article=1166&context=masters_theses **Bagemihl, B. (2000),** *Biological Exuberance: Animal Homosexuality and Natural Diversity*, Stonewall Inn Editions **Osborne, H. (2017),** More Gay Dolphins Observed Off Coast of Western Australia: https://www.newsweek.com/gay-dolphins-australia-homosexual-behavior-645360 **Brennan, P.L.R. et al. (2022),** Evidence of a functional clitoris in dolphins: https://www.cell.com/current-biology/fulltext/S0960-9822(21)01544-X **Orbach, D.N. et al. (2017),** Genital interactions during simulated copulation among marine mammals: https://royalsocietypublishing.org/doi/10.1098/rspb.2017.1265

Common pill woodlouse

Charlat, S. et al. (2003), Evolutionary consequences of Wolbachia infections: https://www.cell.com/trends/genetics/fulltext/S0168-9525(03)00024-6 **Staaf, D. (2023),** Animal Sex Determination Is Weirder Than You Think: https://nautil.us/animal-sex-determination-is-weirder-than-you-think-296080/ **Hillary, V. Edwin and Ceasar, S. Antony (2021),** Chapter Seven – Genome engineering in insects for the control of vector borne diseases: https://www.sciencedirect.com/science/article/abs/pii/S1877117320301873 **Huigens, M.E. et al. (2004),** Natural interspecific and intraspecific horizontal transfer of parthenogenesis-inducing *Wolbachia* in Trichogramma wasps: https://www.ncbi.nlm.nih.gov/pmc/articles/PMC1691627/ **Leclercq, S. et al. (2016),** Birth of a W sex chromosome by horizontal transfer of *Wolbachia* bacterial symbiont genome: https://www.pnas.org/doi/full/10.1073/pnas.1608979113 **Legrand, J.J. et al. (1984),** Nouvelles données sur le déterminisme génétique et épigénétique de la monogénie chez le crustacé isopode terrestre *Armadillidium vulgare* Latr: https://pubmed.ncbi.nlm.nih.gov/22879150/ **Stouthamer, R. (2009),** Chapter 269 – Wolbachia: https://www.sciencedirect.com/science/article/abs/pii/B9780123741448002782

Bluegill sunfish

Cordero, A. et al. (1998), Mating opportunities and mating costs are reduced in androchrome female damselflies, *Ischnura elegans* (Odonata): https://www.sciencedirect.com/science/article/abs/pii/S0003347297906035?via%3Dihub **Ebert, J. (2005),** Cuttlefish win mates with transvestite antics: https://www.nature.com/articles/news050117-9 **Gross, M.R. and Charnov, E.L. (1980),** Alternative male life histories in bluegill sunfish: https://www.ncbi.nlm.nih.gov/pmc/articles/

PMC350407/ **Lamichhaney, S. et al. (2015),** Structural genomic changes underlie alternative reproductive strategies in the ruff (*Philomachus pugnax*): https://www.nature.com/articles/ng.3430 **Neff, B.D. et al. (2003),** Sperm investment and alternative mating tactics in bluegill sunfish (*Lepomis macrochirus*): https://academic.oup.com/beheco/article/14/5/634/186409 **Roughgarden, J. (2004),** *Evolution's Rainbow: Diversity, Gender, and Sexuality in Nature and People*, University of California Press **Shuster, S.M. (1987),** Alternative Reproductive Behaviors: Three Discrete Male Morphs in Paracerceis sculpta, an Intertidal Isopod from the Northern Gulf of California: https://www.jstor.org/stable/1548612 **Whiting, M.J, et al. (2009),** Flat lizard female mimics use sexual deception in visual but not chemical signals: https://www.ncbi.nlm.nih.gov/pmc/articles/PMC2660994/

Common pheasant

Brooks, R. (2022), Bounds of diversity: queer zoology in Europe from Aristotle to John Hunter: https://academic.oup.com/zoolinnean/article/195/1/1/6568055 **Brooks, R. (2019),** Queer Birds: Avian Sex Reversal & the Origins of Modern Sexology: https://www.bshs.org.uk/wp-content/uploads/Viewpoint_119_web_v3.pdf **Gilfillan, G.D. et al. (2016),** Rare observation of the existence and masculine behaviour of maned lionesses in the Okavango Delta, Botswana: https://onlinelibrary.wiley.com/doi/abs/10.1111/aje.12360 **Hunter, J. (1780),** Account of an extraordinary pheasant: https://royalsocietypublishing.org/doi/10.1098/rstl.1780.0030 **Major, A.T. and Smith, C.A. (2016),** Sex Reversal in Birds: https://pubmed.ncbi.nlm.nih.gov/27529790/ **Maryland Department of Natural Resources,** Antlers Tell Much about Deer: https://dnr.maryland.gov/wildlife/pages/hunt_trap/deer_antlers.aspx **Smirnov, A.F. et al. (2022),** Natural and Experimental Sex Reversal in Birds and Other Groups of Vertebrates, with the Exception of Mammals: https://link.springer.com/article/10.1134/S1022795422060114 **Zhang, X. et al. (2023),** Overview of Avian Sex Reversal: https://www.ncbi.nlm.nih.gov/pmc/articles/PMC10079413/#:~:text=Sex%2Dreversed%20birds%20can%20be,reversal%20%5B%211%2C212%5D.

Splitgill mushroom

Briggs, H. (2020), How one teaspoon of Amazon soil teems with fungal life: https://www.bbc.co.uk/news/science-environment-53197650 **Hyde, K.D. (2022),** The numbers of fungi: https://link.springer.com/article/10.1007/s13225-022-00507-y **Kothe, E. (1999),** Mating Types and Pheromone Recognition in the Homobasidiomycete *Schizophyllum commune*: https://www.sciencedirect.com/science/abs/pii/S1087184599911295?via%3Dihub **Scharping, N. (2017),** Why This Fungus Has Over 20,000 Sexes: https://www.discovermagazine.com/planet-earth/why-this-fungus-has-over-20-000-sexes **Whitehouse, H.L.K. (1949),** Hererothallism and sex in the fungi: https://onlinelibrary.wiley.com/doi/abs/10.1111/j.1469-185X.1949.tb00582.x **Wilson, A.M. et al. (2015),** Homothallism: an umbrella term for describing diverse sexual behaviours: https://imafungus.biomedcentral.com/articles/10.5598/imafungus.2015.06.01.13

Chinese shell ginger

Barrett, S.C.H. (2002), The evolution of plant sexual diversity: https://www.nature.com/articles/nrg776 **Barriault, I. et al. (2009),** Flowering period, thermogenesis, and pattern of visiting insects in *Arisaema triphyllum* (Araceae) in Quebec: http://www.aroid.org/gallery/gibernau/2009p/Arisaema%20flowering%20cycle%20-%20Barriault%20et%20al%202009.pdf **Barriault, I. et al. (2010),** Pollination ecology and reproductive success in Jack-in-the-pulpit (*Arisaema triphyllum*) in Québec (Canada): https://pubmed.ncbi.nlm.nih.gov/20653899/ **Li, Q-J. et al. (2001),** Flexible style that encourages outcrossing: https://www.nature.com/articles/35068635 **Li, Q-J. et al. (2002),** Mating system and stigmatic behaviour during flowering of *Alpinia kwangsiensis* (Zingiberaceae): http://sourcedb.xtbg.cas.cn/zw/lw/200908/P020091117580413687588.pdf **Rust, R.W. (1980),** Pollen Movement and Reproduction in *Arisaema triphyllum*: https://www.jstor.org/stable/2484085 **Vitt, P. (1997),** Functional ecology of gender change in *Arisaema triphyllum*: An interdisciplinary approach: https://www.proquest.com/openview/9147ad56b851af1d0b529d3ed3fcd913/1?pq-origsite=gscholar&cbl=18750&diss=y

Cane toad

Barrionuevo, F.J. et al. (2004), Testis-like development of gonads in female moles. New insights on mammalian gonad organogenesis: https://www.sciencedirect.com/science/article/pii/S0012160603007899 **Brown, F.D. et al. (2003),** Bidder's organ in the toad *Bufo marinus*: Effects of orchidectomy on the morphology and expression of lamina-associated polypeptide 2: https://onlinelibrary.wiley.com/doi/full/10.1046/j.1440-169X.2002.00665.x **Carmona, F.D. et al. (2007)**, The evolution of female mole ovotestes evidences high plasticity of mammalian gonad development: https://onlinelibrary.wiley.com/doi/10.1002/jez.b.21209 **Dias, A.B. et al. (2022)**, Bilateral asymmetry in bullfrog testes and fat bodies: correlations with steroidogenic activity, mast cells number and structural proteins: https://pubmed.ncbi.nlm.nih.gov/35287007/ **Jiménez, R, et al. (2022)**, The Biology and Evolution of Fierce Females (Moles and Hyenas): https://www.annualreviews.org/doi/full/10.1146/annurev-animal-050622-043424 **Lambert, M.R. et al. (2019)**, Molecular evidence for sex reversal in wild populations of green frogs (*Rana clamitans*): https://www.ncbi.nlm.nih.gov/pmc/articles/PMC6369831/ **Lambert, M.R. et al. (2015)**, Suburbanization, estrogen contamination, and sex ratio in wild amphibian populations: https://www.ncbi.nlm.nih.gov/pmc/articles/PMC4586825/ **Skelly, D.K. et al. (2010)**, Intersex frogs concentrated in suburban and urban landscapes: https://pubmed.ncbi.nlm.nih.gov/20862600/ **Traijitt, T. et al. (2023)**, Steroidogenic potential of the gonad during sex differentiation in the rice field frog *Hoplobatrachus rugulosus* (Anura: Dicroglossidae): https://onlinelibrary.wiley.com/doi/abs/10.1002/jez.2723 **Whitworth, D.J. et al. (1999)**, Testis-like steroidogenesis in the ovotestis of the European mole, *Talpa europaea*: https://pubmed.ncbi.nlm.nih.gov/9916009/

Moss mites
Brandt, A. et al. (2017), Effective purifying selection in ancient asexual oribatid mites: https://www.nature.com/articles/s41467-017-01002-8 **Brandt, A. et al. (2021)**, Haplotype divergence supports long-term asexuality in the oribatid mite *Oppiella nova*: https://www.pnas.org/doi/10.1073/pnas.2101485118 **Maraun, M. et al. (2019)**, Parthenogenetic vs. sexual reproduction in oribatid mite communities: https://www.ncbi.nlm.nih.gov/pmc/articles/PMC6662391/ **Milius, S. (2000)**, Bdelloids: No sex for over 40 million years: https://www.thefreelibrary.com/Bdelloids%3a+No+sex+for+over+40+million+years.-a062685144 **Pachl, P. et al. (2021)**, Repeated convergent evolution of parthenogenesis in Acariformes (Acari): https://www.ncbi.nlm.nih.gov/pmc/articles/PMC7790623/ **Schaefer, I. and Caruso, T. (2019)**, Oribatid mites show that soil food web complexity and close aboveground-belowground linkages emerged in the early Paleozoic: https://www.ncbi.nlm.nih.gov/pmc/articles/PMC6805910/ **Simon, J-C. et al. (2003)**, Phylogenetic relationships between parthenogens and their sexual relatives: the possible routes to parthenogenesis in animals: https://academic.oup.com/biolinnean/article/79/1/151/2639789?login=false **Stelzer, C-P. et al. (2010)**, Loss of

Sexual Reproduction and Dwarfing in a Small Metazoan: https://www.ncbi.nlm.nih.gov/pmc/articles/PMC2942836/ **Stelzer, C-P. (2007)**, Obligate asex in a rotifer and the role of sexual signals: https://onlinelibrary.wiley.com/doi/10.1111/j.1420-9101.2007.01437.x **Vakhrusheva, O.A. et al. (2020)**, Genomic signatures of recombination in a natural population of the bdelloid rotifer *Adineta vaga*: https://www.ncbi.nlm.nih.gov/pmc/articles/PMC7749112/

Dungowan bush tomato
Albeck-Ripka, L. (2019), Meet Australia's New Sex-Changing Tomato: *Solanum plastisexum*: https://www.nytimes.com/2019/06/18/world/australia/tomato-sex-nonbinary.html **Gebhardt, C. (2016)**, The historical role of species from the Solanaceae plant family in genetic research: https://www.ncbi.nlm.nih.gov/pmc/articles/PMC5121179 **McDonnell, A.J. et al. (2019)**, *Solanum plastisexum*, an enigmatic new bush tomato from the Australian Monsoon Tropics exhibiting breeding system fluidity: https://www.ncbi.nlm.nih.gov/pmc/articles/PMC6592974/ **Pensoft Publisher (2019)**, Scientists challenge notion of binary sexuality with naming of new plant species: https://www.eurekalert.org/news-releases/786165 **Subramaniam, B. and Bartlett, M. (2023)**, Re-imagining Reproduction: The Queer Possibilities of Plants: https://academic.oup.com/icb/article/63/4/946/7110401 **Wooller, S. (2019)**, This tomato is the first 'sexually fluid' plant: https://nypost.com/2019/06/19/this-tomato-is-the-first-sexually-fluid-plant/

Barklice
Cepelewicz, J. (2019), Why Evolution Reversed These Insects' Sex Organs: https://www.quantamagazine.org/why-evolution-reversed-these-insects-sex-organs-20190130/ **Yong, E. (2014)**, In This Insect, Females Have Penises And Males Have Vaginas: https://www.nationalgeographic.com/science/article/in-this-insect-females-have-penises-and-males-have-vaginas **Yong, E. (2017)**, This Common Butterfly Has an Extraordinary Sex Life: https://www.theatlantic.com/science/archive/2017/06/butterfly-cabbage-white-vagina-dentata/530889/ **Yoshizawa, K. et al. (2014)**, Female Penis, Male Vagina, and Their Correlated Evolution in a Cave Insect: https://www.sciencedirect.com/science/article/pii/S0960982214003145?via%3Dihub **Yoshizawa, K. et al. (2018)**, A biological switching valve evolved in the female of a sex-role reversed cave insect to receive multiple seminal packages: https://elifesciences.org/articles/39563 **Yoshizawa, K. et al. (2018)**, Independent origins of female penis and its coevolution with male vagina in cave insects (Psocodea: Prionoglarididae): https://royalsocietypublishing.org/doi/10.1098/rsbl.2018.0533

Picture credits